T0341145

Recent Advances in Distillery Waste Management for Environmental Safety

Recent Advances in Distillery Waste Management for Environmental Safety

VINEET KUMAR
PANKAJ CHOWDHARY
MAULIN P. SHAH

CRC Press
Taylor & Francis Group
Boca Raton London New York

CRC Press is an imprint of the
Taylor & Francis Group, an **informa** business

First edition published 2022
by CRC Press
6000 Broken Sound Parkway NW, Suite 300, Boca Raton, FL 33487-2742

and by CRC Press
2 Park Square, Milton Park, Abingdon, Oxon, OX14 4RN

© 2022 Taylor & Francis Group, LLC

CRC Press is an imprint of Taylor & Francis Group, LLC

Library of Congress Cataloging-in-Publication Data

Names: Kumar, Vineet (Vineet Kumar Rudra), author. | Chowdhary, Pankaj, author. | Shah, Maulin P, author. Title: Recent advances in distillery waste management for environmental safety / authored by Vineet Kumar, Pankaj Chowdhary and Maulin P. Shah.
Description: First edition. | Boca Raton : CRC Press, 2021. | Includes bibliographical references and index.Identifiers: LCCN 2021011183 | ISBN 9780367466015 (hbf) | ISBN 9781032047942 (pbk) | ISBN 9781003029885 (ebk) Subjects: LCSH: Distilling industries—Waste disposal. | Alcohol industry—By-products. | Green chemistry. Classification: LCC TD899.D5 K86 2021 | DDC 338.4/76635—dc23
LC record available at https://lccn.loc.gov/2021011183

ISBN: 978-0-367-46601-5 (hbk)
ISBN: 978-1-003-02988-5 (ebk)
ISBN: 978-1-032-04794-2 (pbk)

DOI: 10.1201/9781003029885

Typeset in Palatino
by KnowledgeWorks Global Ltd.

This book is truly dedicated to our families for their abundant support, patience, understanding, endless love, and educating us to date. Without them this book would not have been possible.

Contents

Preface

Global industrialization over the past century has resulted in the widespread generation of huge quantities of hazardous inorganic and organic wastes. The management and safe disposal of wastes from industrial sources in an economically and environmentally acceptable manner is becoming a serious issue facing modern industry throughout the world. The discharge of huge quantities of waste, both solid (sludge) and liquid (effluent), from alcohol distilleries into the environment creates a threat to all organisms that causes big environmental hassles and health hazards. However, distillery industries are growing rapidly throughout the world due to the potent application of alcohol in the medicinal, cosmetics, food, biochemical, and chemical sectors. The increasing numbers of distilleries in the world have resulted in a substantial increase in industrial waste. The Central Pollution Control Board (CPCB), Government of India, has listed distilleries among the top 17 most polluting industries of India due to their huge effluent volume generation and presence of recalcitrant organic and inorganic pollutants. Thus, the safe disposal and appropriate management and utilization of hazardous waste discharges by distilleries present not only a challenge for the country but also a threat to the scientific society as well.

Over the years, several technologies based on physical, chemical, and thermal processes have been employed for the treatment and management of distillery waste. However, these treatment and management processes offer an economic non-viability, have limited versatility, are labor intensive, have operational constraints, offer partial treatment, may result in the plausible formation of secondary hazardous by-products, and also generate large amounts of toxic sludge, limiting their industrial applicability. In addition, these practices can also destroy waste-degrading microbial communities. Therefore, environmentalists and government bodies are looking for cheap, efficient, and long-lasting sustainable solutions for the management of hazardous distillery waste for its recycling and reuse. Bioremediation, using potent organisms such as fungi, bacteria, actinomycetes, plants, and/or their enzymes, is considered a sustainable, cost-effective, and eco-friendly technology for management of distillery waste in the environment. This technology is gaining importance day by day because it is cheap, feasible, and safe to clean the contaminated localities.

This book, *Recent Advances in Distillery Waste Management for Environmental Safety*, comprising 15 chapters, gives an overview of the latest research and development with in-depth coverage of recent advances in various aspects of biodegradation and bioremediation of distillery waste management for environmental safety. The chapters in this book cover numerous topics, including colorants of distillery waste, phytoremediation, vermifiltration, microbial fuel cell technology, and biodiesel production for distillery waste management. The current knowledge regarding the colorant-persistent organic and inorganic pollutants discharged in distillery waste and their impact on the environment and health hazards are described in detail. Further, this book also presents a brief overview of various approaches (physicochemical and biological) employed for the treatment and management of distillery waste at industrial as well as laboratory scales worldwide. The biodegradation of recalcitrant organic pollutants, especially melanoidins, present in huge quantities in distillery waste is a major challenge for sustainable development in the current scenario. Hence, this book emphasizes the role of ligninolytic enzymes in the degradation and detoxification of melanoidins containing distillery waste. This book has given emphasis to the role of different bioreactors for the treatment of complex distillery waste. The mechanism of phytoremediation of distillery effluent, sludge, as well as contaminated soil and water, is still not very clear to all researchers and academicians. Therefore, the current advances in phytoremediation technology for the remediation of distillery waste have been included in this book. The role of vermifiltration technology for recycling and reuse of hazardous distillery waste has also been described. Further, this book has also given special emphasis to vermifiltration technology for recycling and reuse of distillery waste for environmental protection. Furthermore, the book has imbibed the special emphasis on two-stage sequential treatment approaches as a novel approach for biodegradation and detoxification of distillery wastewater. The book has also covered adequate knowledge of bioelectricity generation and biodiesel production from distillery waste using microbial fuel-cell technology as a new sustainable technique and too for management of distillery waste. Finally, the emerging issues, challenges, and future prospects for the safe disposal and management of distillery waste in the environment after anaerobic treatment have also been highlighted in this book. All chapters in this book are comprehensible and straightforward, with relevant graphical and photographic illustrations to make the topic simple and provide additional help to clarify.

This book can serve as a reference book for those who are interested in knowing about the role of microbes in biodegradation and management of distillery waste. This book will be useful for both novices and experts in the field of bioenergy generation. This book will also be of great value to academicians, researchers, environmentalists, industrialists, professional engineers, policymakers, industry persons, and students at the bachelor's, master's, and doctoral levels, as well as to other enthusiastic people who are wholeheartedly devoted to conserving the environment for sustainable development.

Vineet Kumar
Bilaspur, Chhattisgarh, India

Pankaj Chowdhary
Lucknow, UP, India

Maulin P. Shah
Bharuch, Gujarat, India

Acknowledgments

Recent Advances in Distillery Waste Management for Environmental Safety is the outcome of our end-less efforts of almost three years that we took to complete this book. In this endeavor, we, the authors, were not alone, but assisted by many people. Thus, we are deeply indebted to the many individuals who helped directly or indirectly in the accomplishment of this work with their support, valuable guidance, and innumerable suggestions, many of whom deserve special mention.

First and foremost, we would like to acknowledge and thank our families without whose patience, understanding, and forbearance this book could not have been written. They wholeheartedly supported us in our overburdened schedule and stood with us during this long journey. Any success that we have achieved or will achieve in the future would not be possible without the love and moral support of our beloved families.

We are grateful to Dr. Ram Chandra, Professor, and Dr. R.N. Bharagava, Assistant Professor in the Department of Environmental Microbiology at Babasaheb Bhimrao Ambedkar (A Central) University, Lucknow, Uttar Pradesh, India, for their tremendous academic support and wonderful opportunities that they provided to Dr. Vineet Kumar and Dr. Pankaj Chowdhary during the course of doctoral studies and, most prominently, supporting us in this unique field of metagenomics, phytoremediation, biore-mediation, and biodegradation of industrial waste. We also record our appreciation to the many experts in Bioremediation, Biodegradation, and Waste Management, most particularly Dr. Vinod Kumar Garg, Professor and Dean in the School of Environment and Earth Sciences at Central University of Punjab, Punjab and Dr. Sunil Kumar, Senior Principal Scientist and Head in the Water Reprocessing Division at CSIR-National Environmental Engineering Research Institute, Nagpur, Maharashtra, India, for their outstanding and invaluable advice on how to construct an effective book. They helped us in editing some of the manuscripts for value addition.

We would like to thank those colleagues who took the time to read over individual chapters of this book, and those who reviewed the entire manuscript. Their comments have been gratefully received, and in some cases spared us from the embarrassment of seeing our mistakes perpetuated in print.

We wish to warmly and gratefully acknowledge the editorial and production staff of CRC Press (Taylor & Francis Group) for their excellent work. The team of Taylor & Francis Group has played a great role throughout, always helpful and supportive. Special thanks go out to Dr. Renu Upadhyay, Senior Acquisition/Commissioning Editor (Life Science) for the execution of the publishing agreement, encouragement, support, and valuable suggestions, and to Ms. Jyotsna Jangra, Editorial Assistant (Life Science) at CRC Press (Taylor & Francis Group), India, for constant critical advice and invaluable sup-port throughout the project. Special recognition and sincere appreciation go to Paul Boyd in his role as Production Editor at CRC Press (Taylor & Francis Group), who masterfully managed the book's schedule and progress, keeping communication lines open and ensuring the highest quality at every stage, and to Rajiv Kumar and his team at Knowledge Works Global Ltd. (KWGL), Chennai, India. The authors are extremely grateful to Rajiv Kumar, Project Manager, KWGL, for transforming literally more than 350 pages of text and art manuscript into the superb learning tool you have in front of you. We are also deeply indebted to the many publishing professionals at CRC Press, Taylor & Francis Group, USA and India, for the consistent encouragement, hard work, and careful attention to the development and bringing of this valuable book into the world.

We are grateful to those publishers and individuals who have granted permission to reproduce diagrams.

Finally, Dr. Kumar would like to express his gratitude to his lover, Ms. Priyanka Yadav, for allowing him to devote so many weekends to "the book."

Vineet Kumar

Pankaj Chowdhary

Maulin P. Shah

About the Authors

Dr. Vineet Kumar is presently working as an Assistant Professor in the Department of Botany at Guru Ghasidas Vishwavidyalaya (GGV), Bilaspur, India, and teaches Industrial and Environmental Microbiology and Cell and Molecular Biology at the same institution. Before his joining, he worked as an Assistant Professor and Academic Coordinator at the Vinayak Vidyapeeth, Meerut, India. He obtained his M.Sc. (2010) and M.Phil. (2011) in Microbiology from Ch. Charan Singh University, Meerut, Uttar Pradesh (UP), India. Subsequently, he joined the Department of Environmental Microbiology in 2012 at Babasaheb Bhimrao Ambedkar University (BBAU), UP, India, where he completed his doctoral work on the topic *"Study of bacterial communities in two step treatment of post-methanated distillery effluent by bacteria and constructed wetland plant treatment system."* His research work had been supported by the University Grants Commission. He has also received merit certificates for the best academic contribution by Vice-Chancellor of BBAU. After completion of his doctoral work in 2018, he joined the Department of Microbiology at Dr. Shakuntla Mishra National Rehabilitation University, Lucknow, UP, India, as Guest Faculty, where he has taught courses in general microbiology, microbial genetics, molecular biology, and environmental microbiology. He then went to Jawaharlal Nehru University, New Delhi, India, where he was trained on a DBT-sponsored research project under the supervision of Prof. Indu Shekhar Thakur. His research interests include the exploration of efficient microbe and plant-based ecofriendly and sustainable strategies for the biodegradation, bioremediation, and phytoremediation of environmental pollutants from contaminated sites. Presently, his research activities are focused on production of biodiesel and recovery of bioplastic from industrial waste. Dr. Kumar has also served as a potential reviewer for various scientific journals in his research areas. He has presented several papers relevant to his research areas at national and international conferences. He received a Young Scientist award in 2018 for his excellent contribution to Environmental Microbiology. To date, he has published 14 peer-reviewed research and review articles in high-impact journals, six edited books, and two authored books, and has authored/co-authored more than 32 chapters in edited books. In addition, he has also authored/co-authored four research articles published in conference proceedings, and four magazine articles on different aspects of bioremediation and phytoremediation of industrial waste pollutants. He is an active member of various scientific societies, including the Indian Science Congress Association (ISCA), Kolkata, India, the Association of Microbiologists of India (AMI), Lucknow, and the Biotech Research Society (BRSI), India. He is the founder of the Society for Green Environment, India (www.sgeindia.org). Outside the lab, Dr. Kumar enjoys bicycling, gardening, traveling, spending time with family, and reading spiritual books. He can be reached at drvineet.micro@gmail.com; vineet.way18@gmail.com.

Dr. Pankaj Chowdhary is President of the Society for Green Environment (SGE) and currently he is working as a Postdoctoral Fellow in the Environmental Toxicology Group at the Indian Institute of Toxicology Research, Lucknow, India. He obtained his Ph.D. (2018) in the area of Microbiology from the Department of Environmental Microbiology at Babasaheb Bhimrao Ambedkar University (A Central University), Lucknow, Uttar Pradesh, India. During his Ph.D., his work mainly focused on the role of ligninolytic enzyme-producing bacterial strains in the decolorizing and degradation of coloring compounds from distillery wastewater. His main research areas are Microbial Biotechnology,

Biodegradation and Bioremediation of Environmental contaminants in industrial wastewaters, and Metagenomics. Currently, he is working on the synthesis of biochar using various types of lignocellulosic waste. He has edited two international books entitled *Emerging and Eco-Friendly Approaches for Waste Management* and *Microorganisms for Sustainable Environment and Health.* He has published many research/review papers in national and international peer-reviewed journals of high-impact factor published by Springer, Elsevier, Royal Society of Chemistry (RSC), Taylor & Francis Group, and Frontiers. He has also published many national and international book chapters and magazine articles on the biodegradation and bioremediation of industrial pollutants. He has presented many posters/papers relevant to his research areas in national and international conferences. He is a life member of the Association of Microbiologists of India (AMI) and the Indian Science Congress Association (ISCA), Kolkata, India.

 **Dr. Maulin P. Shah** has been an active researcher and scientific writer in his field for over 22 years. He received his Ph.D. (2009) in Microbiology from Sardar Patel University, Vallabh Vidyanagar (Gujarat), India. His research interests include biological wastewater treatment, environmental microbiology, biodegradation, bioremediation, and phytoremediation of environmental pollutants from industrial wastewaters. He has published more than 250 research articles in national and international journals of repute on various aspects of microbial biodegradation and bioremediation of environmental pollutants. He has edited 52 books published by Elsevier, Springer, CRC Press, RSC, and De Gruyter. He has presented several papers relevant to his research areas in national and international conferences. He has also been serving as a regular reviewer for various scientific journals in his research areas. He is the Founder Editor-in-Chief of the *International Journal of Environmental Bioremediation & Biodegradation* (Science and Education Publishing, USA; from 2011 to 2014) and the *Journal of Applied & Environmental Microbiology* (Science and Education Publishing, USA; from 2011 to 2014). He is the Editor-in-Chief of the *Journal of Advances in Biotechnology (JBT).* He is also the Editor and Associate Editor of many scientific journals in his field. He is also serving as a member of the Editorial Board of the more than 200 scientific journals published from the reputed publisher.

1

Introduction

1.1 Brief Background

The worldwide demand for energy, uncertainty of natural resources, and concern about global warming have led to the environment-friendly development of alternative liquid biofuels. Ethanol is regarded as one of the excellent candidates since it reduces dependence on fossil-fuel reserves and it is also cleaner burning and thus better for air quality. India is the fourth largest producer of ethanol after Brazil, the United States, and China, and the 5% blending of petrol/motor fuel is mandatory all over the country, which helps in reducing import of crude oil, thereby saving foreign exchange. Currently, the 5% blending is applicable only in ten states and three union territories and it requires about 410 million liters of anhydrous ethanol. Apart from its use in petro fuel, beverages, medicines, pharmaceuticals, and flavoring compounds, ethanol is an important feedstock for the manufacture of various chemicals like acetic acid, butanol, butadiene, acetic anhydride, polyvinyl chloride, etc., which are being used in the production of rubber, drugs, solvents, and pesticides. Due to its high demand in global market, distilleries are growing at an alarming rate in the world (Kumar and Sharma 2019). However, distilleries are widely regarded as one of the most polluting industries in more than 130 countries, especially in developing countries including India, Mexico, Brazil, and Japan, as 88% of its raw materials are converted into waste and discharged as a large volume of high-strength effluent (Kumar and Chandra 2020a). According to one estimate, for every liter of ethanol that is produced during sugarcane-molasses-based fermentation and distillation processes, about 10–15 L of recalcitrant, troublesome, and complex liquid as an effluent, also called spent wash, stillage, vinasses, mosto, and raw distillery effluent, is generated (Chandra and Kumar 2017a). Sugarcane molasses, a natural sweetener obtained as a by-product during the processing of refined sugar from sugarcane (*Saccharum officinarum*) juice, contain 45–50% of residual sugars (i.e., glucose, fructose, and sucrose), 15–20% of non-sugar organic substances, 10–15% of ash (minerals), and about 20% of water. The majority of distilleries are located in tropical and subtropical regions of the world using sugarcane molasses as a feedstock (Chandra and Kumar 2017b). About 90% of the molasses produced in cane sugar manufacture is consumed in ethanol production. Moreover, some distilleries are using various substrates as a feedstock such as cereal malt (i.e., rice, barley, wheat, and maize) and grapes for ethanol fermentation. It has been reported that spent wash produced in sugarcane-molasses-based distilleries has a high organic load as compared to other raw material used for ethanol production (Chandra et al. 2018a). India has a large network of distilleries of varying capacity that are distributed throughout the country (Kumar and Chandra 2018a). A recent report suggests that there are 397 molasses-based distilleries in India producing 8,679 million liters of alcohol and generating 3.5×10^{13} kL of spent wash as a liquid waste annually (Kumar and Chandra 2020b). Depending on the sugarcane origin and the subsequent fermentation and distillation processes for ethanol production and waste treatment, the volume and intrinsic composition of generated high-strength spent wash can vary significantly (Kumar 2021). Spent wash is a dark brownish color effluent characterized by a specific obnoxious odor, high ash content, high biological oxygen demand (BOD), chemical oxygen demand (COD), total organic carbon (TOC), total dissolved solids (TDS), total soluble solids (TSS), and organic matter and refractory organic compounds such as a variety of sugar decomposition products, phenolics, steroids, anthocyanins, tannins, furfurans, melanoidins, androgenic-mutagenic compounds, and endocrine-disrupting compounds (EDCs), that are highly toxic in nature and resistant to biodegradation in an open environment (Chandra and Kumar 2017a; Chandra et al. 2018a). In addition, a high mineral load was also reported due to the presence of sulfate (SO_4^{2-}), potassium (K^+), phosphorous (PO_4^{2-}), calcium (Ca^{2+}), and sodium (Na^+). Moreover, spent wash also contains a large concentration of numerous heavy metals (HMs) such as iron (Fe), zinc (Zn), nickel (Ni), manganese (Mn), lead (Pb), mercury (Hg), copper (Cu), and chromium (Cr) (Kumar et al. 2020a). The intense color in spent wash is mainly due to the existence of a dark brown polymeric pigment compound known as melanoidin, which is formed by Maillard reaction (MR), a non-enzymatic browning reaction between amino and sugar compounds. Melanoidin possesses antioxidant, antimicrobial, and antihypertensive activity properties. Therefore, the presence of these antimicrobial compounds and the removal of color in spent wash together pose a major challenge to scientists, environmentalist, and researchers for efficient and sustainable development (Kumar and Sharma 2019). It has been reported that melanoidins have net negative charges; hence, different heavy metallic ions such as Cu^{2+}, Cr^{6+}, Cd^{2+}, Fe^{2+}, Zn^{2+}, and Ni^{2+} strongly bind with melanoidins to form organometallic complex in distillery spent wash (Hatano et al. 2016; Kumar and Chandra 2020b). A few adequate uses for spent wash management have been identified and it is used in large-scale operations, such as recycling to fermentation streams, energy production, animal feed production, and ferti-irrigation practices, i.e., utilizing it as a liquid fertilizer for sustainable agriculture and reducing the water input for plant growth (Kumar and Chopra 2012; Kumari et al. 2016). However, ferti-irrigation practices usually have negative effects on soil and groundwater quality in long term due to accumulation of their low pH and organic and inorganic contents (Kumari et al. 2009; Kumari et al. 2012). Furthermore, some studies have indicated that spent wash also negatively affects the microbial flora of soil (Chowdhary et al. 2018a). In addition to spent wash generation during ethanol production, a huge amount of solid waste as a yeast sludge is formed in the distilleries, that cause pollution when it is disposed into the environment without adequate treatment. However, yeast sludge is rich in protein and contains a considerable amount of essential amino acids, and drying sludge grains are marketed as livestock feed and make it the best source for the production of single-cell protein. Figure 1.1 outlines the ethanol production processes, generation of liquid and solid waste and their treatment by different physicochemical, integrated, and biological approaches, and impact of discharged wastes on the environment.

The organic and inorganic pollutants present in the effluent of different nature are reacting to each other and make the effluent more toxic and complex (Zhang et al. 2017; Kumar and Chandra 2020b). Due to high pollution nature of spent wash, the Ministry of Environment, Forest, and Climate Change (MoEF&CC), Government of India, has listed ethanol industries at the top among the "Red Category" industries. In accordance with the environmental protection act and rules of the MoEF&CC and Central Pollution Control Board (CPCB), Government of India, it is mandatory for distilleries to treat hazardous spent wash before it is disposed into the environment. According to Indian government rules, every industry has an effluent treatment plant. However, the plants are not generally operated because of the high cost involved in treating effluents; as a result, they discharge the untreated effluent to the outside environment, which ultimately negatively affects human beings. In many instances, industries dilute the spent wash by mixing with raw water before discharge in order to meet the set waste disposal standard. This dilution, even though accepted in some regions, is of great environmental concern as it does not reduce the absolute pollution load of the spent wash. Thus, Indian government's policies on pollution prevention have forced distilleries to look for cost-effective and sustainable technology for decreasing the characteristics of the final discharged spent wash.

Distillery spent wash treatment is being carried out generally by four routes in the industry: (i) concentration followed by incineration, (ii) direct oxidation by air at a high temperature followed by aerobic treatment, (iii) anaerobic digestion with biogas recovery followed by aerobic polishing, and (iv) reverse osmosis. Out of four routes, anaerobic digestion, also known as biomethanation, is the most attractive and the best possible treatment method for decolorization and detoxification of spent wash in developing countries such as Japan, India, and Mexico, due to its low-cost, easy operation, and eco-friendly technique besides its energy generation potential in the form of biogas (biomethane, CH_4), which is utilized for running steam boilers for the generation of electricity (Malik et al. 2019a). The effluent received from distilleries after anaerobic digestion, called as biomethanated spent wash or

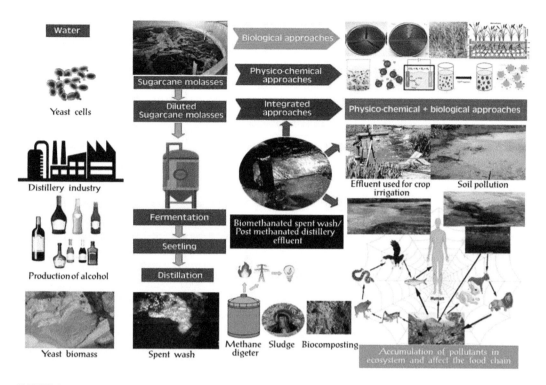

FIGURE 1.1 The production of ethanol from fermentation of sugarcane-molasses, generation of effluent and sludge as a waste by-products and their treatment by various physicochemical, biological, and integrated approaches, and impact of disposed wastes into the environment.

biomethanated distillery effluent, is required to further cope with environmental standards (Mohana et al. 2007; Ravikumar et al. 2011; Reis et al. 2019). Biomethanated spent wash is characterized by an extremely higher level of BOD, COD, TDS, total Kjeldahl nitrogen (TKN), SO_3^{2-}, Na^+, and PO_4^{3-}, with alkaline pH and dark brown color (Saner et al. 2014; Shinde et al. 2020). Moreover, biomethanated spent wash retains a high amount of various toxic metals, namely, Ni, Mn, Fe, Zn, Cu, Pb, and Cd, along with melanoidins and various recalcitrant organic pollutants (ROPs) (Singh et al. 2010; Sharma et al. 2011; Wagh and Nemade 2018).

Since the anaerobic digestion is reported to remove about 40–50% COD, 60–65% BOD, and color of spent wash converts darker with higher TDS after anaerobic digestion due to complexation of organic and inorganic pollutants (Chandra et al. 2018a). This means that spent wash after anaerobic digestion still contains some organic load and is not safe for discharge into the environment (Kumar et al. 2021a). In India, the existing full-scale distillery effluent treatment system includes the combination of anaerobic digestion and a two-stage extended aeration process. Although, the primary aerobic and anaerobic conventional treatment systems can easily remove sugars, volatile organics, and other easily biodegradable compounds that constitute BOD, the color and COD constituted by melanoidin pigment remain unchanged after treatment due to its antioxidant and antimicrobial nature, and these treatments are found ineffective to degrade the color compounds at high concentrations. Besides effluent, sugarcane-molasses-based distilleries also produce a huge amount of solid sludge as a by-product during anaerobic digestion of spent wash, which contributes significantly toward the contamination of environments (Chandra and Kumar 2017b, c; Mahaly et al. 2018). Different approaches to distillery sludge management such as incineration, landfilling, and composting are reported in scientific literature by a various group of researchers (Suthar and Singh 2008; Singh et al. 2014). Some research works evaluated the potential of composting of distillery sludge, also mixed with other substrates, for its utilization in agriculture as a fertilizer; the main concern related to this option is the fate of heavy metals and organic chemicals contained in distillery sludge and their effects on plant growth and soil quality. Sludge discharged after anaerobic digestion of spent wash enriched with a high concentration of Fe and other metals have adverse effects on the environment (Chandra and Kumar 2017c; Kumar and Chandra 2020b). Plants growing in the metalliferous soils may have the ability to cope with high metal concentrations. The sludge produced in the anaerobic digestion process is normally used as a substitute of the compost after drying or it is mixed with press mud and then converted into the compost. Preliminary studies conducted in laboratory revealed that the post-methanation bio-sludge of molasses-based distillery is relatively richer in water-soluble organic carbon (2.67%) than other organic manures like farmyard manure (0.20–0.33%) or press mud compost (0.19–0.33%) (Srivastava et al. 2009). The organic matter present in the bio-sludge is similar to the natural fulvic and humic acids. The shortfalls of existing treatment technologies include the huge fuel consumption for evaporation and incineration, power consumption for air diffusion for aerobic oxidation of organic matter, and pollution of groundwater and surrounding lands by leaching of pollutants from aerated lagoon (Krishnamoorthy et al. 2017). However, these methods are expensive and not environment-friendly as they generate large amounts of sludge, which also requires further safe disposal and can also cause secondary pollution (Prasad and Srivastava 2009; Reis et al. 2019). Thus, distillers face the daunting task of effectively treating a high volume of concentrated or partly treated spent wash for its safe disposal into the environment. Spent wash after anaerobic treatment takes a high value as a fertilizer, due to its high organic matter and micronutrient content, that facilitates the growth of crops and is often used in crop fertigation (Ayyasamy et al. 2008; Narain et al. 2012; Kumari et al. 2016). However, only at higher concentration of effluent, it has some adverse impact on the nutrient contents of the soil when used in large quantities. Draining the untreated or inadequately treated and indiscriminately disposed spent wash is causing extensive soil and ground and surface water pollutions (Kumar and Gopal 2001; Sharma et al. 2011; Ravikumar 2015). Moreover, the presence of phenolic compounds in irrigation water represents health and environmental hazard. Thus, proper spent wash treatments are mandatory to remove contaminants before its disposal into the environment (Chandra et al. 2012; España-Gamboa et al. 2017). Considering the strict environmental norms imposed by the Indian government's CPCB on freshwater utilization (maximum consumption of 15 L of freshwater per liter of ethanol production) and zero-liquid discharge from distilleries, alternatives to existing treatment options such as reverse osmosis, incineration, and anaerobic digestion continue to be of interest.

Recent advances in newly developed processes to treat distillery effluent include different physical, chemical, biological, and integrated/sequential techniques (Latif et al. 2011; Singh et al. 2014; Sankaran et al. 2015; Takle et al. 2018). A wide range of physicochemical posttreatment methods and strategies like coagulation/flocculation with aluminum chlorohydrate ($Al_2Cl(OH)_5$), ferric chloride ($FeCl_3$), magnesium chloride ($MgCl_2$), and lime; a low-molecular-weight poly(diallyldimethylammonium chloride) (PDADMAC) (Liang et al. 2009a, b; Fan et al. 2011; Jack et al. 2014; Zhang et al. 2018) adsorption on sugarcane bagasse, peat, fly ash, and zeolite; osmosis with biomimetic membrane (Kumari et al. 2012; Ahmed et al. 2020); oxidation with ozonation/hydrogen peroxide and manganese oxides (Sangave et al. 2007; Singh and Dikshit 2012; Malik et al. 2019a, b); electrocoagulation (Jack et al. 2014), electroflotation (Liakos and Lazaridis 2014), electrochemical (Thakur et al. 2009; Asaithambi et al. 2012), and photocatalyst (photodegradation) processes with aluminum oxide, nanoparticle, and kaolin clay (David and Arivazhagan 2015; Mabuza et al. 2017); Fenton and photo-Fenton-based advanced oxidation processes (Lima et al. 2006; Arimi 2017); and filtration (membrane filtration, ultrafiltration, nanofiltration, sand filtration) (Nataraj et al. 2006; Rai et al. 2008; Pant and Adholeya 2009a; Rafigh and Rahimpour Soleymani 2019), thermal pretreatment, wet air oxidation, concentration-incineration, and combine aerobic/anaerobic, have been suggested or tested for effective elimination of melanoidins and decolorization of untreated or partly treated distillery effluent (Miyata et al. 2000; Apollo and Aoyi 2016; Reis et al. 2019). However, these methods are associated with high operational cost, excess use of chemicals, sensitivity to variable water input, less effective decolorization rate, and a huge amount of toxic sludge generation with subsequent disposal problems (Santal et al. 2011; Kumar et al. 2021b). Moreover, the complete or partial mineralization of hazardous organic molecules leads to the formation of more toxic intermediates and end products. Thus, sufficient treatment is essential before distillery effluent is

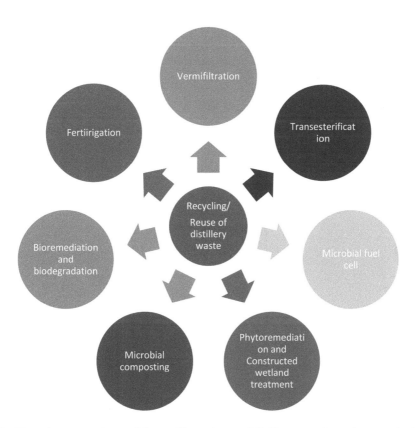

FIGURE 1.2 The various approaches used for recycling and reuse of distillery waste for environmental safety and sustainable development.

disposed into the environment (Kumar et al. 2021c). The ever-increasing generation of effluent from distilleries on the one hand and stringent legislative regulations of its disposal on the other hand have stimulated the need for developing new technologies to process this effluent efficiently and economically. Hence, cost-effective and eco-friendly treatment techniques are urgently required for the efficient treatment of distillery effluent.

In recent years, considerable research efforts in academic, industrial, and government institutions have been focused on the development of innovative biotechnological methods and strategies for the effective treatment of distillery waste as well as to improve the mineralization process of most of the organic pollutants present in the effluent that can recover and reuse effluent for sustainable development. Figure 1.2 highlights the role of various approaches for recycling and reuse of distillery waste for sustainable development. In this context, the total number of articles published by various researchers' groups around the globe on the treatment and management of distillery waste is shown in Figure 1.3.

Due to the relatively low cost and the variations of work progress, the microbe-based bioremediation/ biodegradation of distillery waste has intensified in recent years as mankind strives to find sustainable ways to clean up and restore contaminated environments all over the world.

Bioremediation is an eco-friendly technique that employs many different microbes acting in parallel or sequentially to degrade and detoxify toxic contaminants present in distillery waste (Kumar et al. 2018). Bioremediation is defined as the acceleration of the natural metabolic process whereby microorganisms (i.e., bacteria, actinomycetes, fungi, yeast, cyanobacteria/microalgae), green plant (termed phytoremediation), or their intracellular and/or extracellular enzymes, namely, laccases, lignin peroxidases, and manganese peroxidases, degrade or transform toxic contaminants to carbon dioxide, water, microbial biomass, inorganic salts, and other by-products (metabolites) that may be less toxic than the parent

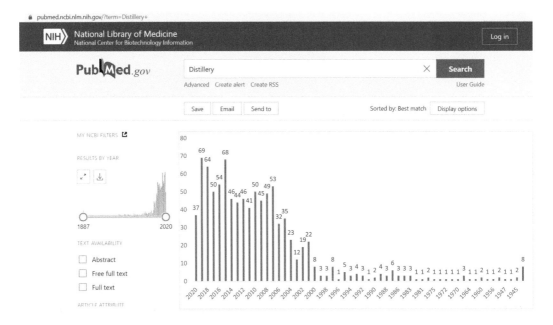

FIGURE 1.3 Articles published on distillery waste management by various groups of researchers around the globe as per the record of PubMed.

compounds. Among them, phytoremediation is a promising, sustainable, and inexpensive approach for the treatment and/or removal of contaminants from the water, wastewater, soil, or sludge or to render them harmless by plants and their associated microbial communities (Hatano et al. 2016; Kumar 2021). Phytoremediation has emerged as an energy-efficient remediation technology for sequestration, detoxification, and mineralization of toxicants through complex natural biological, physiological, and chemical processes and activities of plants and microbes both (Chandra and Kumar 2018b). Depending upon the detoxification process, applicability, medium, type, and extent of pollution, phytoremediation processes can be subdivided into different types such as phytostabilization—the reduction of metal mobility in soil, which accordingly decreases wind-blown dust, minimizes soil erosion, and reduces the contaminant solubility or bioavailability to the food chain; phytodegradation—direct degradation of pollutants by plant enzymes; phytoextraction—mainly uses plants accumulating high concentrations of heavy metals, that can be harvested, discarded, and even extracted to recover metals; rhizodegradation—enhances the activity of telluric degrading microorganisms through the release of root exudates by the plants; phytovolatilization—the release of volatile pollutants to the atmosphere via plant transpiration; rhizofiltration—a root zone in in situ or ex situ technology can be used for the elimination of metals from water and aqueous waste streams that are retained only within the roots of aquatic plants. It reduces the mobility of metals and prevents their migration to the groundwater, thus reducing bioavailability for entry into the food chain. Thus, the development of alternative treatment methods that utilize the advantages of natural processes in the ecosystem is increasing in the area of industrial management. Figure 1.4 illustrates the general overview of generation of solid and liquid wastes by fermentation and distillery industries and their various treatment and management approaches and impact of waste on the aquatic and terrestrial ecosystem.

The aim of this book is to provide a concise discussion on the waste discharged from distilleries, and its major colorants, environmental impact, toxicity profile, and health hazards of numerous organic and inorganic pollutants discharged in distillery effluent. Further, we also describe various treatment approaches using physical, chemical, and biological means, with special emphasis on sequence-based biotechnological approaches as an eco-friendly technique for remediation of contaminated sites and provide a concise discussion on how microbes could be exploited to enhance the phytoremediation efficacy

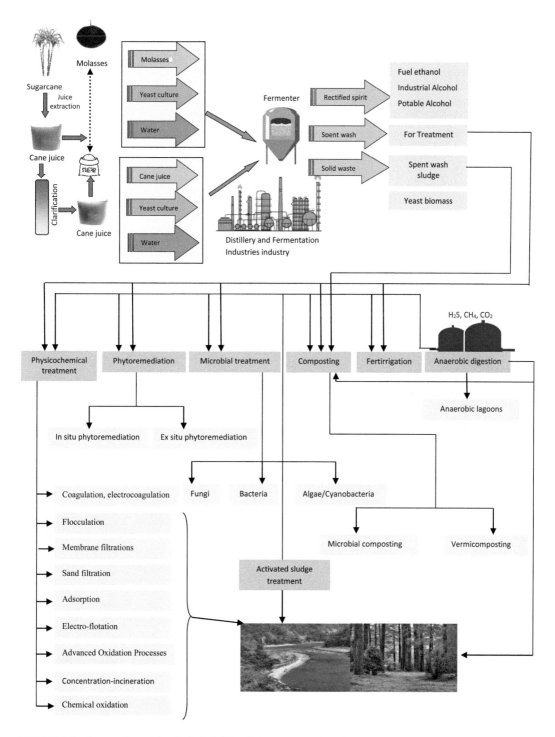

FIGURE 1.4 An overview of the alcohol distillery industry, generation of waste, and their treatment and management approaches.

of plants in contaminated environments. We also discussed the role of ligninolytic enzymes and bioreactors in treatment of recalcitrant distillery effluent. Moreover, the use of plant-microbe-based technology (phytoremediation) for the management of distillery waste is also discussed. Further, we also discussed the role of vermifiltration technology in treatment and management of distillery waste, and we provide some information on costs. Furthermore, the use of a two-step sequential approach for the treatment of distillery effluent is highlighted. Besides, the efficient way of distillery waste for bioelectricity production and biodiesel production as a sustainable approach for waste management is highlighted. Moreover, we also discussed the emerging issues, challenges, future prospects, and developed rules and policies for efficient treatment and management of inorganic and organic compounds containing distillery waste.

2

Distillery Waste Generation and Characteristics

2.1 Introduction

Ethanol-producing distillery industry is currently making a substantial contribution to the world economy on account of its massive potential for employment, growth, and exports in many developing countries. However, there is serious environmental trouble with ethanol production from sugarcane molasses, which is generally connected to the generation of a huge volume of the complex, dark brown-color liquid waste as effluent (Narain et al. 2012; Santal et al. 2016; Ahmed et al. 2020). Distilleries use different kinds of raw material/feedstock such as sugar-based materials (i.e., sugarcane juice and sugar beet molasses), and starch-based substances (i.e., corn, barley, wheat, rice, and cassava), and cellulosic materials (i.e., crop residues, and sugarcane bagasse) for the production of ethanol. Among them, Indian distilleries exclusively use sugarcane-molasses as a most common raw material for the production of ethanol (Kumar and Chandra 2020b). Ogunwole et al. (2020) reported that *Saccharum officinarum* molasses adversely altered testicular and epididymal integrity via lipid peroxidation, thus reducing sperm quality and androgen levels in male Wistar rats. The cycle of raw materials starts from farms, through product formation and fermentation to distillery for the production of alcoholic beverages. Despite the alcohol production, the ethanol-manufacturing process also produces spent wash, yeast sludge, and spent wash sludge as a waste by-product (Chandra and Kumar 2017a). Most of the distilleries are unable to manage such a huge quantum of the waste efficiently and economically. It has been reported that the effluent from molasses-based alcohol distillery has a higher content of organic, inorganic, and organometallic compounds as compared to other raw material used for ethanol production (Tiwari et al. 2012; Tripathy et al. 2020). The high organic load of spent wash is composed of recalcitrant polymeric macromolecules mainly melanoidins that are originated by a non-enzymatic reaction called Maillard reaction (MR) between amino acids and sugar and caramels from overheated sugars that are responsible for their color and odor (Echavarría et al. 2012; Chavan et al. 2013; Wagh and Nemade 2018). The structure of melanoidin and spent wash discharged before and after anaerobic digestion by distilleries are shown in Figure 2.1.

The composition of distillery spent wash varies from industry to industry and from country to country, depending on the manufacturing process, type and quality of raw material used, the equipment used in the factory, and type of treatment technologies adopted. The generated spent wash and overall contribution of melanoidins, sugar, acids, and alcohol in total dissolved solids (TDS) of spent wash is illustrated in Figure 2.2.

As a result of raw distillery spent wash disposal into the ecosystem, a large number of health effects have been occurring to crops, aquatic, terrestrial biota, and humans (Kumar and Gopal 2001; Yadav and Chandra 2011; Jain et al. 2005; Jain and Srivastava 2012). Thus, adequate treatment is warranted before the effluent is discharged into the environment. Anaerobic digestion has been widely considered as the most attractive first step technique for the treatment of spent wash due to its reputation as a low-cost technique besides its biomethane generation potential (Saner et al. 2014; López et al. 2017). Moreover, anaerobic digestion treatment is preferred due to the fact that a great component of spent wash is biodegradable. However, anaerobic digestion is reported to remove about 75–90% of chemical oxygen demand (COD) and 80–90% of biological oxygen demand (BOD). This means after anaerobic digestion, spent wash still retains high COD, BOD and low biodegradability index (BI: BOD_5/COD <0.2), and substantial color imparted by the recalcitrant melanoidins, which restricts the further

DOI: 10.1201/9781003029885-2

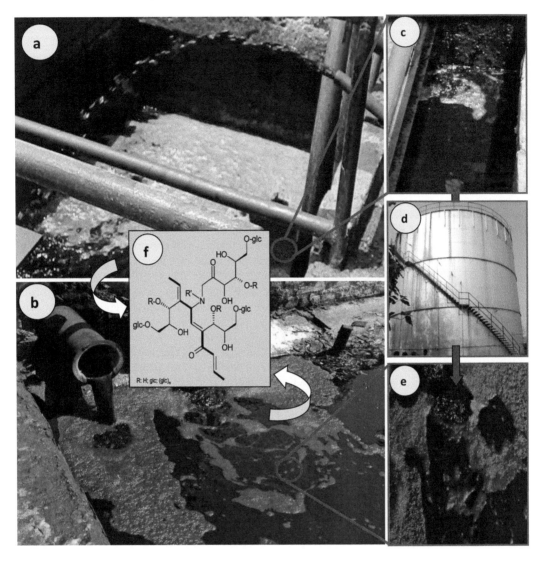

FIGURE 2.1 Colored effluent discharged from distilleries during ethanol production (a) spent wash (b) biomethanated spent wash (c) a large view of the spent wash (d) biomethanated plant (e) a large view of biomethanated spent wash, and (f) melanoidins, a major complex and brown colorant of distillery effluent.

biological treatment of the residual wastewater known as biomethanated spent wash and is not safe for discharge into the environment (Sankaran et al. 2014). The antioxidizing and antimicrobial properties of melanoidins and plant phenolics make the effluent toxic for many organisms, thus, primary biological aerobic and anaerobic treatments have been found ineffective to degrade this color compound. This necessitates appropriate secondary (post)treatment of distillery effluent after the primary biological treatment processes. Therefore, most of the distilleries employed a direct two-stage aerobic process for further treatment of the biomethanated spent wash (Chandra et al. 2012; Junior et al. 2020). Spent wash treatment generates large quantities of sludge as a solid waste that needs to be carefully managed and its safe disposal into the environment represents one of the major problems in distillery industry (Pant and Adholeya 2009b; Kumar and Chandra 2020b). In this chapter, we describe the processes for generation of different wastes in distilleries. Further, the chemical nature of these wastes is also highlighted.

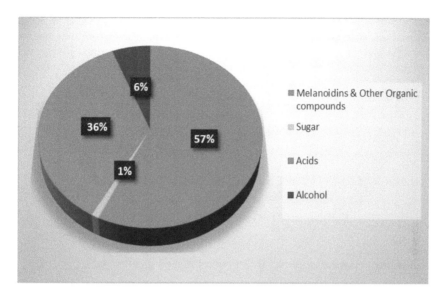

FIGURE 2.2 Contribution of melanoidins, sugar, acids, and alcohol to total dissolved solids (TDS) of distillery effluent (Shinde et al., 2020).

2.2 General Process for Waste Generation in Molasses-Based Distillery

In distilleries, the ethanol manufacturing process broadly consists of three major steps—feed preparation (fermentable sugar-containing diluted molasses solution), fermentation (conversion of sugars to ethanol under anaerobic conditions), and distillation (separation and purification of ethanol).

2.2.1 Feed Preparation

The first manufacturing process for ethanol in a distillery involves the dilution of raw molasses with water, followed by fermentation and distillation. Molasses is the mother-liquor leftover after crystallization of sugar from concentrated cane juice. It is used as chief feedstock/carbon source material in fermentation and distilleries to produce ethanol because it contains a high level of fermentable sugars (45–50%). Thus, it is suitably diluted in order to maintain the desired sucrose level in the range of 15–16%. Moreover, several food supplements of nitrogen and phosphate are added in a fermentation broth.

2.2.2 Fermentation

In this process, the diluted molasses fed to the fermentation tank where it is inoculated with propagated yeast culture (*Saccharomyces cerevisiae*) in about 10:1 proportion, which converts the sugar components of diluted molasses into bioethanol under anaerobic and controlled environmental conditions for a stipulated period of 24–30 h, and the yeast sludge (fermenter sludge) settles down at the end of the process as a waste by-product. In molasses-based distilleries, the fermentation process can be carried out by three modes: (i) batch, (ii) fed-batch, and (iii) continuous mode. In a batch process, molasses is diluted with water to reduce the sugar content and then yeast inoculum is added to this diluted molasses. Further, this diluted molasses is allowed to ferment for 30–40 h. After completion of fermentation, the yeast sludge is separated from the bottom of the fermenter and the fermented wash as the main product of fermentation containing 8–10% ethanol is sent to the analyzer column for distillation where a mixture of steam and bioethanol vapors is collected at the top of the column and the brownish liquid as the bottom product known as spent wash contributes to the pollution load from the distilleries. The fermenter sludge is mainly separated from the bioethanol solution by filtration. In a fed-batch process, a combination of batch and a continuous mode, substrates are supplied to the fermenter during cultivation and the product(s) remain in the bioreactor until the end of the run. Continuous mode is carried out by continually adding culture medium, substrates, and nutrients into a bioreactor containing microorganisms. During this process, the culture volume must be constant and the products formed after fermentation are continuously taken from the media. The fermentation process involves the following steps:

a. Conversion of sucrose to glucose and fructose:

$$\underset{\text{(Sucrose)}}{C_{12}H_{22}O_{11}} + \underset{\text{(Water)}}{H_2O} \xrightarrow{\text{Invertase}} \underset{\text{(Glucose)}}{C_6H_{12}O_6} + \underset{\text{(Fructose)}}{C_6H_{12}O_6} \qquad \text{(i)}$$

b. Fermentation of glucose to alcohol, releasing carbon dioxide and heat:

$$\underset{\text{(Glucose)}}{C_6H_{12}O_6} + \underset{\text{(Water)}}{H_2O} \xrightarrow{\text{Zymase}} \underset{\text{(Ethanol)}}{2C_2H_5OH} + \underset{\text{Carbon dioxide}}{2CO_2} + \text{Heat} \qquad \text{(ii)}$$

Once the fermentation is complete, the cell-free broth is taken for distillation.

2.2.3 Distillation

Distillation proceeds with heating the cell-free fermented broth (fermented wash) to about 90°, and is sent to the degasifying section of the analyzer column. The bubble cap fractionating column removes any trapped gases (CO_2, etc.) from the liquor, which is then steam heated and fractionated to give 40% alcohol. The down-coming dark brown discharge from the analyzer column is known as spent wash. The 40% alcohol stream from the top of the analyzer column is next fed to the bottom of the rectifier column where the temperature is maintained at about 95–100 °C. Water and alcohol vapor gets condensed at different levels in this column and rectified spirit (95%) is withdrawn leaving behind the spent lees. Fuel alcohol or absolute alcohol is produced by the dehydration of rectified spirit through azeotropic distillation or molecular sieve technology.

2.3 Water Uses in Molasses-Based Distilleries

During ethanol production, distillery operations use water for both process and non-process applications.

2.3.1 Process Applications

The process applications include the preparation of sugarcane molasses for fermentation, yeast propagation, and steam requirements for distillation. Freshwater consumption in process applications is in the range of 14.5–21.4 L/L of ethanol production.

2.3.2 Non-Process Applications

The non-process applications involve cooling water uses in making potable alcohol. Water consumption in non-process applications such as cooling water, steam generation, making potable liquor, boiler water, wash water, etc., is much higher, i.e., ranging between 102.65 and 240 L/L of ethanol production.

2.4 Origin and Composition of Distillery Waste

The distillery industry produces a humongous amount of solid as well as liquid waste materials such as (a) yeast sludge from fermenters or molasses sludge from clarifiers meant for the settling of fermented wash, (b) concentrated or high-strength process effluent (commonly known as spent wash) that originates from the analyzer column in distillation process during ethanol production, and (c) spent lees from the rectifier column. Besides, diluted effluent or low-strength process effluent is generated that originates from equipment cleaning (i.e., fermenters and distillation columns) and floor washing. The non-process wastewater is comparatively pure and as such can be recycled. The major sources of wastewater in molasses-based distilleries are fermenter sludge, spent wash, and spent lees. Figure 2.3 gives a simplified description of the generation of fermenter sludge, spent wash, and spent lees from in molasses-based distilleries.

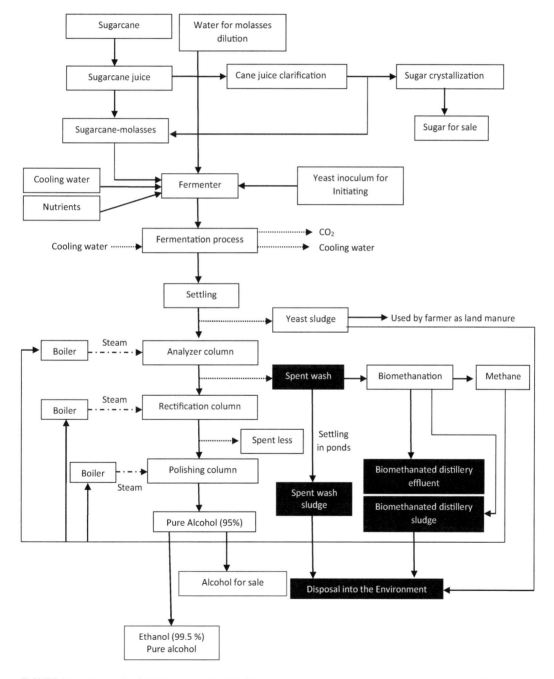

FIGURE 2.3 Schematic of sugar and ethanol manufacturing and generation of waste in molasses-based distillery.

2.4.1 Fermenter Sludge

Once the fermentation is over, agitation in the fermenter due to carbon monooxide generation also subsides and sedimentation of sludge takes place. The settled sludge in the fermenter is discarded and discharged separately, which mainly contains yeast sludge and molasses sludge. The volume of fermenter sludge, commonly called yeast sludge, is about 0.3 L/L of rectified spirit produced. This sludge has a solid content of about 30% by weight, which comprises mostly of the spent yeast and mineral matter. Yeast sludge is rich in protein and contains a considerable amount of essential amino acids and drying sludge grains are marketed as livestock feed and make it the best source for the production of single-cell protein. Fermenter sludge has a higher BOD and a lower volume as compared to spent wash.

Yeast sludge stream is allowed to pass through a series of pits, and sludge is allowed to settle in the pits. The settled sludge is then periodically removed manually and disposed of as it is. The sludge so collected still contains an appreciable quantity of moisture and possess odor nuisance and also handling problems. Yeast is highly proteinous and undergoes biodegradation in anaerobic phase producing various products that are foam-conducive. Continuous yeast ingress into anaerobic reactor is found to generate foam. In addition to this, yeast sludge also exerts a high organic load on anaerobic reactors, resulting in overloading of the treatment system. The spent yeast is highly biodegradable and has BOD of up to 25,000 mg/L. The composition of the yeast sludge generated from a typical batch-type fermentation-based distillery is given in **Table 2.1**.

TABLE 2.1

Typical Characteristics of Raw and Washed Yeast Sludge

Reference	Sharif et al. (2012)				
Country	Pakistan				
Nutrient (%)	**Raw Yeast Sludge**	**Fermenter Washed Sludge**	**Amino Acid Profile (%)**	**Raw Yeast Sludge**	**Fermenter Washed Sludge**
Metabolizable energy (kcal/kg)	2,200	2,375	Lysine	0.435	0.549
Crude protein	27.40	34.80	Methionine	0.542	0.689
True protein	18.10	28.20	Threonine	0.827	1.047
Ether extract	1.10	1.20	Arginine	1.465	1.861
Crude fiber	0.00	0.00	Leucine	1.183	1.503
Ash	22.08	11.88	Isoleucine	1.019	1.294
Acid soluble ash	20.10	7.91	Valine	0.989 b	1.252
Nitrogen free extract	49.52	52.22	Phenylalanine	0.912	1.158
Mineral contents			Histidine	0.504	0.640
Calcium	3.440	1.330	Tyrosine	0.537	0.685
Phosphorus	1.220	0.550	Cystine	1.016	1.291
Potassium	3.520	1.293	Proline	0.619	0.786
Sodium	1.142	0.571	Glutamate	4.553	5.776
Magnesium	2.000	1.200	Serine	0.589	0.751
Manganese	0.010	0.004	Aspartate	1.726	2.192
Iron	0.194	0.107			
Zinc	0.010	0.008			
Copper	0.045	0.033			
Cobalt	0.003	0.002			
Chromium	0.011	0.010			

It is advisable to dewater fermenter sludge and dispose of without mixing it with spent wash as it will increase the BOD of the receiving stream. A lot of solids such as yeast sludge, molasses sludge, and inorganic solids postprecipitated after distillation and entered into spent wash settle at the bottom of the anaerobic reactor and create problems like choking outlet weirs in case of down-low reactors and disturbing sludge blanket reactors also leading to rising in pH, and foaming in reactors. Moreover, dissolved solids in the spent wash precipitates at a higher temperature in the distillation column, which further increases the inorganic sludge content in the spent wash. In continuous fermentation, there is no separate yeast sludge stream as such. However, the fermented wash is generally taken to Lamella clarifiers where yeast sludge and molasses sludge are removed from fermented wash before their feeding to distillation column. Many times, the settling efficiency of these clarifiers is very low due to various reasons and solids are carried over along with supernatant liquid ultimately reaching spent wash. Nature and settling characteristics of yeast sludge are not fully understood/explored, so the best treatment option is yet to be evolved.

2.4.2 Spent Lees

Spent lees are the hot and colorless liquid as a waste obtained from rectifier column at the generation rate of 1.7–2.0 L/L of rectified spirit. It is usually recycled and sometimes mixed with the spent wash or let out separately. BOD/COD values of the spent less are usually low and depend on the alcohol. No specific treatment plant is generally given for spent lees treatment. Recently, many industries are discharging spent lees in the secondary treatment plant.

2.4.3 Spent Wash

The wastewaters discharged from the analyzer column, yeast sludge, spent less, water treatment plant, waste wash water, cooling water, boiler as blowdown, bottling plant, and other wastes as termed as spent wash. Spent wash is the largest and complex effluent stream generated from distilleries in terms of both quality and quantity (Figure 2.4). In distilleries, the major source of wastewater generation is the distillation step wherein a huge volume of dark brown effluent is generated. Spent wash is an undesirable, viscous, hydrophilic, highly acidic (pH from 3.8 to 5.0), deep brown effluent, with a high concentration of organic materials and solids. The volume of spent wash is about 12–15 L/L of rectified spirit produced. Spent wash has many unwanted features: unpleasant odor, high acidity, dark brown color and high TDS, total soluble solids (TSS), COD, BOD, inorganic minerals, and suspended solids (Acharya et al. 2011). **Table 2.2** lists the physicochemical characteristics of distillery wastewaters from all over the world. The high-strength effluent also contains various toxic organic compounds such as butanedioic acid, benzenepropanoic acid, 2-hydroxyisocaproic acid, vanillyl propionic acid, 2-furancarboxylic acid, benzoic acid, and tricarballylic acid and several recalcitrant organic pollutants (ROPs), as presented in **Table 2.3** (Chandra and Kumar 2017a).

Spent wash also contains dissolved impurities from the sugarcane juice, by-products of the fermentation, and nutrients added during fermentation of sugarcane molasses and the breakdown products of various sugars present in sugarcane juice and settled in the bottom of effluent storage tank as solid sludge (Figure 2.4). The suspended impurities such as cellulosic fibers and dust are usually separated before the evaporation of sugarcane juice. However, non-sugars and minerals, water-soluble hemicelluloses, gums, and proteins from the sugarcane juice are present in the spent wash in their converted or original forms, exerting an oxygen demand during subsequent treatment. Practically in the distillery, the spent wash production rates and characteristics are highly variable depending on the type of feedstocks (quality of molasses), fermentation technique, and unit operation in molasses processing and ethanol recovery process. These factors are the major contributors to the pollution load of the distillery effluent. Average spent wash generation is highest in the batch process (11.1–15.0 L/L ethanol production), higher in the continuous process (8.5–11.0 L/L ethanol production), and lowest in the bio-still process (6–8 L/L ethanol production). The presence of toxic/recalcitrant compounds

FIGURE 2.4 Distillery waste (a and b) generated spent wash stored in storage tank and (c) sludge solid settled at the bottom of storage tank (Kumar and Sharma 2019).

TABLE 2.2

Typical Characteristics of Spent Wash Belonging to Different Sources and Countries as Reported by Different Investigators

References	Chandra and Kumar (2017a)	–	Sankaran et al. (2014)	Fito et al. (2018)	Thiyagu and Sivarajan (2018)	Rath et al. (2010)
Country	**India**		**India**	**Ethiopia**	**India**	**India**
Parameters						
Color appearance	Dark brown	–	Dark brown	–	Dark brown	–
Temperature (°C)	–	–	80–90	–	–	–
pH	4.07	3.0–4.5	4.0–4.6	3.9	3.8	7.23
Color (Pt–Co)	150,000	–	–	–	–	–
COD (mg/L)	90,000	110,000–190,000	85,000–110,000 ppm	132,445	162,000	30,520
BOD$_5$ (mg/L)	42,000	50,000–60,000	25,000–35,000 ppm	40,271	–	15,300
EC (µS/cm)	–	–	26–31	–	45.5	28,700
TS (mg/L)	83,084	110,000–190,000	–	150,300	140,260	35,340

TABLE 2.2 *(Continued)*

References	Chandra and Kumar (2017a)	–	Sankaran et al. (2014)	Fito et al. (2018)	Thiyagu and Sivarajan (2018)	Rath et al. (2010)
Country	India		India	Ethiopia	India	India
TVS (mg/L)	–	80,000–120,000	–	–	–	–
TSS (mg/L)	–	13,000–15,000	4,500–7,000 ppm	–	27,860	9,980
TDS (mg/L)	77,776	90,000–150,000	85,000–110,000 ppm	–	112,400	27,240
Sodium (mg/L)				207.6–263.0	–	420
Chlorides (mg/L)	2,200	8,000–8,500	4,500–8,400 ppm	6722	10,650	5,626
Phenols (mg/L)	–	8,000–10,000	3,000–4,000 ppm	–	–	–
Calcium (mg/L)				1,787.4–3,389.8	2,975	920
TKN (mg/L)	–	–	–	–	–	636.25
TN (mg/L)	2,800	5,000–7,000	4,200–4,800 ppm	–	–	–
NO₃-N (mg/L)	–	–	–	3.2	–	–
Na₂CO₃ (mg/L)	–	–	–	–	–	–
Phosphate (mg/L)	5.36	2,500–2,700	1,500–2,200 ppm	21.2	2,100	–
Sulfates (mg/L)	5,760	7,500–9,000	13,100–13,800 ppm	4,502	3,015	5,100
Sulfides (mg/L)	–	–	–	–	–	–
Zink (mg/L)	2.487	–	–	2.8	–	1.09
Nickel (mg/L)	1.175	–	–	2.7	–	0.145
Manganese (mg/L)	4.556	–	–	6.6	–	1429
Iron (mg/L)	163.947	–	–	19.6	–	6.3
Lead (mg/L)	BDL	–	–	–	–	0.19
Chromium (mg/L)	1.05	–	–	2.3	–	0.067
Copper (mg/L)	0.337	–	–	1.5	–	0.265
Cadmium (mg/L)	BDL	–	–	–	–	0.036
Magnesium (mg/L)	–	–	–	927.6	2,380	753.25

Note: Some of these real effluent characteristics are not within the typical range of values presented in the table.

COD: chemical oxygen demand; BOD₅: 5-day biochemical oxygen demand; TOC: total organic carbon; EC: electrical conductivity; TS: total solids; TSS: total suspended solids; TDS: total dissolved solids; TVS: total volatile solids; TA: total alkalinity; TKN: total Kjeldahl nitrogen; TN: total nitrogen; NO₃-N: nitrate-nitrogen; Na₂CO₃: sodium carbonate; NaOH: sodium hydroxide; NaCl: sodium chloride; BDL: Below detection limit

TABLE 2.3

List of the Organic Compounds Detected by GC-MS after Derivatization of (a) Acetone, (b) Isopropanol, (c) Ethyl Acetate and (d) Methanol Extract Derived from Spent Wash (Chandra and Kumar 2017a)

(a)		(c)	
S. No.	**Organic Compounds**	**S. No.**	**Organic Compounds**
1.	1,3-Propanediol, TMS ether	1.	L-Lactic acid, TMS ether, TMS ester
2.	Propanoic acid, 3-[(TMS)oxy)-TMS ether	2.	Acetic acid, [(TMS)oxy], TMS ester
3.	Butanoic acid, 3-methyl-2-[(TMS)oxy]-TMS ether	3.	Butanoic acid, 3-methyl-2-[(TMS), TMS ester
4.	D-Erythrotetrofuranose, tris-*O*-(TMS)	4.	Valeric acid, 5-methoxy, TMS ester
5.	Pentanoic acid	5.	2-Hydroxysocaproic acid, TMS ether
6.	Butanedioic acid, bis(TMS) ester	6.	Ethyl(TMS) succinate
7.	Resorcinol, *O*-bis(TMS)	7.	1,3-Propanediol, TMS ether
8.	2,3-Butandiol, bis-*O*-(TMS)	8.	Butanedioic acid, bis(TMS) ester
9.	Malic acid (*O*-(trimethylsilyl)-bis (trimethylsilyl ester)	9.	Diethyl methylsuccinate
10.	2-Methyl-1,3-butanediol 2TMS	10.	Lactic acid dimer, bis(TMS)

(Continued)

TABLE 2.3   (*Continued*)

(a)		(c)	
S. No.	**Organic Compounds**	**S. No.**	**Organic Compounds**
11.	2-Furancarboxylic acid,5-[[(TMS) oxy] methyl], TMS ester	11.	Silane, [1,4-phenylenebis(oxy)]bis(trimethyl)
12.	2,3,5-Tri-*O*-TMS-arabino-1,5-lactone	12.	2-Methyl-1,3-propanediol 2TMS
13.	Cyclooctene, 1,2-bis(trimethylsiloxy)	13.	2-Furancarboxylic acid, 5-[[(trimethylsilyl)oxy] methyl], TMS ester
14.	Tricarballylic acid TMS	14.	Benzenepropanoic acid, α-[(TMS)oxy], TMS ester
15.	D-Ribo-hexanoic acid, 3-deoxy-2,56, tris-*O*-(TMS), lactone	15.	Benzoic acid, 4-[(TMS)oxy], TMS ester
16.	Benzoic acid, 3,4-bis[(TMS)oxy], TMS ester	16.	Ethyl-TMS dipropylmalonate
17.	*tert*-Butylhydroquinone, bis(TMS) ether	17.	Bicyclo[4.3.0]nonane-2-one,[*Z*]-*cis*-8-(phenyl-1-trimethylsilylmethylene)
18.	Trimethylsilyl 3,5-dimethoxy-4-9TMS oxy) benzoate	18.	2-Hydroxyheptanoic acid 2TMS
19.	Vanillypropionic acid, bis(TMS)	19.	D-Erythro-Hex-2-enoic acid, 2,3,-di-*O*-methyl-5,6-bis-*O*-(TMS)-Y-lactone
20.	Benzeneacetic acid, α,4-bis[(TMS) oxy]-methyl ester	20.	Tricarballylic acid 3TMS
		21.	Benzoic acid, 3-methoxy-4-[(TMS)oxy], TMS ester
(b)	**Organic Compounds**	22.	Trimethylsilyl 3,5 dimethoxy-4-(TMS oxy) benzoate
1.	Propanoic acid, 2–9 (TMS)oxy],-TMS ester	23.	3-Vanil-1,2-propanediol 3TMS
2.	Butanedioic acid, bis(TMS) ester	24.	β-D-Galactopyranoside, methyl 2,6-bis-*O*-(TMS) cyclic butyboronate
3.	Butane, 1,2,4-tris(trimethylsilyloxy)	**(d)**	**Organic Compounds**
4.	2-Methyl-1,3-propanediol-2-TMS	1.	Propanoic acid, 3-[(TMS)oxy], TMS ester
5.	Compounds ethanol	2.	Butanoic acid, 3-methyl-2-[(TMS) oxy], TMS ester
6.	Ethyl(trimethyl) succinate	3.	2-Methylbutanoic acid, 3-(*t*-butyldimethylsilyloxy), methyl ester
7.	Butanedioic acid, bis(TMS) ester	4.	Erythro-pentitol, 2-dedoxy-1,3,4,5-tetrakis-*O*-(TMS)
8.	3,6-Dioxa-2,7-disilaoctane,2,2,4,7,7-pentamethyl	5.	3,6-Didoxa-2,7-disilaoctane, 2,2,4,5,7,7-hexamethyl
9.	Erythritol per-TMS 2,3,4,5-tetrahydroxypentanoic acid-1,4-lactone, tris(TMS)	6.	Cyclooctene, 1,2-bis-(trimethylsilyloxy)
10.	Cyclooctene, 1,2-bis(trimethylsiloxy)	7.	2,3,5,-Tri-*O*-trimethylsily-arabino-1,5-lactone
11.	D-Ribo-hexanoic acid, 3-deoxy-2,5,6-tris-*O*-(TMS) lactone	8.	D-Ribo-hexanoic acid, 3-deoxy-2,5,6 tris-*O*-(TMS)-lactone
12.	α-D-Galactopyranose, 1,2,3-tris-*O*-(TMS), cyclic methylboronate		
13.	D-Lactic acid-Di TMS		
14.	*n*-Hexane extract of spent wash		
15.	Benzene, 1-ethyl-3,5-disopropyl		
16.	Eicosane		
17.	3,4-Dihyroxymandelic acid, ethyl ester, tri-TMS		
18.	Octadecane, 3-ethyl-5(2-ethylbutyl)		

represents a challenge for its efficient treatment, especially in those regions where expensive technologies cannot be easily adopted. The acidic pH (3.8–5) of spent wash is associated with the presence of organic acids produced by the yeasts during the alcoholic fermentation process and causes the dissolution of metal ions in the water bodies. Moreover, spent wash possesses a high concentration of reducing sugars, hemicelluloses, lignin, polysaccharides, phenolics, anthocyanins, tannins, fatty acids, sterols, and resins. Distillery spent wash also harbored a bacterial community, which seemed to just thrive in the spent wash without notable multiplication and proliferation (Chandra and Kumar 2017a). Mohana et al. (2007) indicated that bacterial community present in the spent wash could not be able to utilize the carbon compounds present in the spent wash for its growth and multiplication. However, supplementation of diluted spent wash with M63 medium containing glucose as carbon source was able to show bacterial growth. The lack of bacterial growth in the spent wash may be either because of the deficiency of the nutritional ingredients or because of the presence of toxic levels of pollutants. The dilution of the spent wash is performed in order to reduce the level of toxic ingredients, which might be present in the spent wash. The rapid advances in molecular techniques have opened the possibilities to elucidate microbial community structure, physiological functions, and genetic interactions in nature. The major pollutant types identified in spent wash are summarized in **Table 2.3**.

The unpleasant odor of spent wash is due to the existence of skatole, indole, and other sulfur-containing compounds. Moreover, the dark brown color of spent wash is imparted by compounds, namely, melanoidins, caramel, hexose alkaline degradation products (HADP), furfurans (from acid hydrolysis), lignin, polyphenols, and plant pigments such as carotenoids, chlorophyll, anthocyanins, tannins, which make spent wash more complex and recalcitrant. These colorants are concentrated in molasses after the crystallization of sugar and are further transferred to the spent slurry during the fermentation of molasses.

2.4.4 Anaerobically Digested Spent Wash/Biomethanated Distillery Effluent

Biomethanated distillery effluent (BMDE) is a residual liquid color waste generated after anaerobic digestion of spent wash (Figure 2.5). It is characterized by high values in terms of several physicochemical and biological parameters, indicating the degree of pollution, including coloring, temperature, salinity, pH, BOD, COD, TDS, total nitrogen (TN), total phosphorus (TP), and non-biodegradable organic compounds (Mohana et al. 2007); on the other hand, these effluents also contain heavy metal ions such as zinc (Zn), nickel (Ni), manganese (Mn), iron (Fe), lead (Pb), chromium (Cr), copper (Cu), and cadmium (Cd) (Table 2.3) and residual reducing sugars, phenolics, lipids, proteins, amino acids, organic acids and various recalcitrant organic compounds (**Tables 2.4 and 2.5**; Chandra et al. 2018a).

According to the United States Environmental Protection Agency (EPA 2012), most of these organic compounds listed in **Table 2.4** are toxic, carcinogenic, mutagenic, and endocrine disruptors in nature and some organic compounds in the effluents are resistant to biodegradation. During the treatment processes, it is important to monitor and compare these parameters with the standard concentrations before discharging the corresponding effluent to the receiving water body. Monitoring of the treatment performance regarding other parameters such as total organic carbon (TOC), ammonia-nitrogen (NH_4^+-N), nitrate-nitrogen (NO_3-N), and *ortho*-phosphate-phosphorus (PO_4-P) is also required. The major pollutants in biomethanated distillery effluent are organic and inorganic chemicals such as phenol, Maillard reaction products (MRPs), ADPH, endocrine-disrupting chemicals (EDCs), salts, total phosphate, dissolved solids, suspended solids, total solids, and heavy metals. Coloring matters (i.e., MRPs) are the major contaminant in the BMDE and has to be removed before discharging the effluent into the environment. Without proper treatment, the colored effluent creates an aesthetic problem and its color discourages the downstream use of wastewater. The removal of MRPs from effluent has been rated to be relatively more important, which usually contribute to the major fraction of BOD. Accordingly, distillery effluent may contain chemicals that are toxic, carcinogenic, mutagenic, or teratogenic to various aquatic and terrestrial organisms.

FIGURE 2.5   Distillery waste (a) Biomethanated spent wash in storage tank (b) Sludge solid settled at the bottom of storage tank (c) Biomethanated effluent sludge discharged after anaerobic digestion of spent wash.

2.4.5 Anaerobically Digested Sludge

Distillery is one of the highly polluting industries due to discharge of the huge amount of sludge as a by-product during anaerobic digestion of spent wash (Figure 2.5). Anaerobically digested distillery sludge is considered as a source of toxic heavy metals and various androgenic-mutagenic compounds into the environment, and its disposal in the environment is problematic (**Table 2.6**). Besides, distillery sludge also consists of a mixture of several recalcitrant organic compounds along with melanoidins, as listed in **Table 2.7**. Thus, sludge must be treated and reused or disposed of to ensure environmental protection and maximum benefits. The spent wash sludge contains a huge amount of water (more than 90%) along with organic solids, which cause problems in its transportation, treatment, and disposal. Therefore, sludge volume reduction is important to minimize the operating, treatment, and disposal costs. Sludge incineration, settling, dewatering, and degradation are important processes for spent wash sludge recycling and disposal

TABLE 2.4

Typical Characteristics of Anaerobically Digested Spent Wash Belonging to Different Sources and Countries as Reported by Different Investigators

References	Sankaran et al. (2014)	Zhang et al. (2017)	Zhang et al. (2017)	Yadav and Chandra (2012)	Mohana et al. (2007)	Singh and Sharma (2012)
Country	India	–	–	India	India	India
Type of wastewater	Anaerobically Treated Molasses	Cassava Raw stillage	Cassava Raw stillage	Biomethanated Spent Wash	Anaerobically Treated Spent Wash	Biomethanated Spent Wash
Parameters						
Temperature (°C)	35–40	–	–	–	–	–
pH (–)	7.5–8.0	3.9–4.5	7.4–7.5	8.20	7.5–8	7.8
Color	Dark brown	Dark brown	Dark brown	–	–	–
Color (Pt-Co)	–	2,200–2,950	1250–2,245	35,000	–	112,500
Turbidity (NTU)	40	–	–	–	–	–
EC (µS/cm)	31–36 m	6.3–6.7	5.4–5.9	–	–	32.7
BOD (mg/L)	7,000–10,000 ppm	–	928–1,422	15,200	8,000–10,000	18,500
COD (mg/L)	25,000–40,000 ppm	33,870–40,400	1,650–2,552	36,000	45,000–52,000	54,000
TS (mg/L)	–	–	–	39,496	72,500	–
TSS (mg/L)	22,000–34,000 ppm	–	–	26,842	40,700	5,200
TDS (mg/L)	35,000–45,000 ppm	–	–	12,654	7,997	23,100
Chlorides (mg/L)	–	–	–	750	–	2,600
Sodium (mg/L)	–	–	–	420	–	–
Phenols (mg/L)	–	–	–	180	7,202	850
TKN (mg/L)	–	–	–	–	–	–
TN (mg/L)	–	680–900	260–533	–	4,284	–
NO$_3$-N (mg/L)	350–400 ppm	–	–	–	–	–
Ammonical nitrogen (NH$_4^+$-N)	1,000–1,100 ppm	–	–	–	–	–
CaCO$_3$ (mg/L)	600 ppm	–	–	–	–	–
Calcium (mg/L)	–	–	–	1,433	1,625	1,717
Na$_2$CO$_3$ (mg/L)	–	–	–	–	–	–
Phosphate (mg/L)	400 ppm	–	–	–	–	–
Potassium (mg/L)	–	–	–	54	–	–

(*Continued*)

TABLE 2.4 *(Continued)*

References	Sankaran et al. (2014)	Zhang et al. (2017)	Zhang et al. (2017)	Yadav and Chandra (2012)	Mohana et al. (2007)	Singh and Sharma (2012)
Country	India	—	—	India	India	India
Type of wastewater	Anaerobically Treated Molasses	Cassava Raw stillage	Cassava Raw stillage	Biomethanated Spent Wash	Anaerobically Treated Spent Wash	Biomethanated Spent Wash
Parameters						
Sulfates (mg/L)	4,000–4,500 ppm	—	—	2,457	3,875	—
Total sugar (g%)	—	—	—	—	0.36	—
Volatile fatty acids (mg/L)	—	—	—	—	—	27,720
Reducing sugar (g%)	—	—	—	—	0.17	—
Iron (mg/L)	10 ppm	—	—	—	—	—
Zinc (mg/L)	—	—	—	10.52	—	—
Copper (mg/L)	—	—	—	1.64	—	—
Cadmium (mg/L)	—	—	—	0.013	—	—
Nickel (mg/L)	—	—	—	—	—	—
Manganese (mg/L)	—	—	—	40.70	—	—
Iron (mg/L)	—	—	—	—	—	—
Arsenic (mg/L)	—	—	—	—	—	—
Magnesium (mg/L)	—	—	—	—	—	1717

Note: Some of these real effluent characteristics are not within the typical range of values given in the table.

COD: chemical oxygen demand; BOD$_5$: five-day biochemical oxygen demand; TOC: total organic carbon; EC: electrical conductivity; TS: total solids; TSS: total suspended solids; TDS: total dissolved solids; TVS: total volatile solids; TA: total alkalinity; TKN: total Kjeldahl nitrogen; TN: total nitrogen; NO$_3$-N: nitrate-nitrogen; Na$_2$CO$_3$: sodium carbonate; NaOH: sodium hydroxide; NaCl: sodium chloride.

TABLE 2.5

List of the Organic Compounds Detected by GC-MS after Derivatization of the Ethyl Acetate Extract Derived from Melanoidins Containing Distillery Effluent (Chandra et al. 2018a)

S. No.	Organic Compounds	S. No.	Organic Compounds
1.	Silanol, trimethyl-, trimester with boric acid	18.	11-*cis*-Octadecenoic acid, TMS ester
2.	L-Lactic acid	19.	9,12-Octadecanoic acid(Z,Z)-, TMS ester
3.	Bis(dimethyl-*t*-butylsilyl) oxalate	20.	Octadecanoic acid
4.	Cyclohexanol, 4-[(TMS)oxy]-*cis*	21.	1-Eicosanol
5.	D-(−)Lactic acid, trimethyl ether, TMS ester	22.	1,2-Benzenedicarboxylic acid, bis(2-ethylhexyl) ester
6.	Ethanedioic acid	23.	1,2-Benzenedicarboxylic acid, mono(2-ethylhexyl) ester
7.	*t*-Butyldimethyl(2-styry[1,3]dithian-2-yl) silane	24.	Hexadecanoic acid
8.	Butane-1,3-diol, 1-methylene-3-methyl, bis(TMS) ether	25.	Hexadecanoic acid, 2,3-bis[(TMS)oxy]propyl ester
9.	Silanol, trimethyl-,benzoate	26.	1,7-Pentatriacontene
10.	Propane, 1,2,3-tris[(*tert*-butyldimethylsilyl)oxy]	27.	β-Sitosterol
11.	Acetic acid, [bis[(TMS)oxy]phosphinyl]-TMS ester	28.	Querecetin 7, 3′,4′ trimethoxy
12.	Pyrrolozine1-one,7-propyl	29.	2-Monostearin
13.	Undecenoic acid	30.	1-Monolinoleoylglycerol TMS ester
14.	Dodecanoic acid	31.	Octadecanoic acid, 2,3 bis[(TMS)oxy]propyl ester
15.	Dotriacontane	32.	Octadecanoic acid, ethyl ester
16.	Pyrrolo(1,2-*a*)pyrazine-1,4-dione,	33.	Stigmasta-5, 22-dien-3-ol (3β,22E)
17.	Hexadecanoic acid	34.	Silane[[(3β)-cholesta-5-en-3-yl]trimethyl

TABLE 2.6

Typical Characteristics of Anaerobically Digested Distillery Sludge Belonging to Different Distilleries Located in Different Countries as Reported by Different Group of Researchers

References	Chandra and Kumar (2017c)	Chandra and Kumar (2017b)	Chandra et al. (2018b)	Kumar and Chandra (2020b)	Permissible Discharge Limit (USEPA 2002)
Parameters					
pH	8.0	8.0	8.1	8.12	–
EC (µS/cm)	2.292	4.1	4.12	4.5	–
Chlorides (mg/L)	1824.4	–	–	–	–
NH_4^+-N (mg/L)	190	190	–	15.70	–
TN (mg/L)		2.463	–	3.56	1
NO_3-N (mg/L)	110	110	85.89	–	–
Sodium	56.16	56	42.13	39.11	200
Chloride	–	1,825	1,272.74	891.14	1,500
Phosphate (mg/L)	–	–	2,268.83	1,827.23	–
Total organic carbon	–	17.318	–	12.21	–
Total hydrogen	–	4.013	–	3.12	–
Total oxygen	–	36.251	–	32.16	–
Sulfates (mg/L)	–	–	145.07	135.07	–
Zink (mg/L)	210.624	210.15	43.47	94.25	2.0
Nickel (mg/L)	13.425	13.425	15.60	10.115	0.1
Manganese (mg/L)	126.292	126.30	238.47	95.15	0.20
Iron (mg/L)	2,403.64	2,403	5,264.49	1,512	2.0

(Continued)

TABLE 2.6 (*Continued*)

References	Chandra and Kumar (2017c)	Chandra and Kumar (2017b)	Chandra et al. (2018b)	Kumar and Chandra (2020b)	Permissible Discharge Limit (USEPA 2002)
Parameters					
Copper (Cu) (mg/L)	73.638	73.62	847.46	61.55	0.5
Chromium (Cr) (mg/L)	21.847	21.825	–	17.524	0.05
Cadmium (Cd) (mg/L)	1.446	1.440	–	1.012	0.01
Lead (Pb) (mg/L)	16.332	16.33	31.22	–	0.05

COD: chemical oxygen demand; BOD-5: five-day biochemical oxygen demand; TOC: total organic carbon; EC: electrical conductivity; TS: total solids; TSS: total suspended solids; TDS: total dissolved solids; TVS: total volatile solids; TA: total alkalinity; TKN: total Kjeldahl nitrogen; TN: total nitrogen; NO_3-N: nitrate-nitrogen; Na_2CO_3: sodium carbonate; NaOH: sodium hydroxide; NaCl: sodium chloride

TABLE 2.7

List of the Organic Compounds Detected by GC-MS after Derivatization of (a and b) *n*-hexane, (c and d) Ethyl Acetate Extracts Derived from Post-Methanated Distillery Sludge as Reported by Different Investigators (Chandra and Kumar 2017b; Chandra Et Al. 2018b; Kumar and Chandra 2020b)

(a)		(c)	
S. No.	**Organic Compounds**	**S. No.**	**Organic Compounds**
1.	2-Methyl-4-keto pentan-2-OL	1.	Silane, (4-ethylphenyl)trimethyl
2.	D-Lactic acid, TMS ether, TMS ester	2.	Benzene, 1-ethyl-2-methyl
3.	1-Methylene-3-methyl-butanol	3.	2,3-D-2-Methylsuccinic acid 2TMS
4.	Benzene, 1,3-bis(1,1-dimethylethyl)	4.	Ethanedioic acid, bis(TMS) ester
5.	Phosphoric acid	5.	β-Eudesmol, TMS ether
6.	2-Isoropyl-5-methyl-1-heptanol	6.	Benzoic acid, 2-methyl-, TMS ester
7.	1-Phenyl-1-propanol	7.	Phenol, 2,4-bis(1,1-dimethylethyl)
8.	Tetradecane	8.	Phenol, 2,6-bis(1,1-dimethylethyl)
9.	Decane, 2,3,5,8-tetramethyl	9.	Tetradecanoic acid, TMS ester
10.	Propanoic acid	10.	9,12-Octadecadienoic acid (Z,Z)-TMS ester
11.	1-Dodecanol	11.	Benzoic acid, 3,4,5 tris(TMS oxy)-TMS ester
12.	Docosane	12.	11-*trans*-Octadecenoic acid, TMS ester
13.	Dodecanoic acid	13.	*cis*-10-Nonadecenoic acid, TMS ester
14.	*tert*-Hexadecanethiol	14.	Hexadecanoic acid, 2-hydroxy-1-(hydroxymethyl) ethyl ester
15.	Tetradecanoic acid	15.	Hexadecanoic acid, 2,3-bis[(TMS)oxy]propyl ester
16.	*n*-Pentadecanoic acid	16.	Docosanoic acid, TMS ester
17.	Hexadecanoic acid, TMS ester	17.	2-Monostearin TMS ether
18.	Octadecanoic acid	18.	Octadecanoic acid, 2,3-bis[(TMS)oxy]propyl ester
19.	2,6,10,14,18,22-Tetracosahexane 2,6,10,18,19,23-hexamethyl	19.	Dotriacontane
20.	Stigmasta-5,22-dien-3-ol(3b,22E)	20.	Hexacosanoic acid
21.	Stigmasterol	21.	Silane,[[(3β,22E)-ergosta-7,22-dien-3-yl]trimethyl
22.	Lanosta-8, 24 dien-3-one	22.	Octacosanol
23.	Spirostan-3-one (5*a*, 20*b*, 25*R*)	23.	Stigmasterol TMS ether
24.	β-Sitosterol trimethyl ether	24.	5α-Cholestane,4-methylene
		25.	24-Ethyl-δ-(22)-coprostenol, TMS

TABLE 2.7 *(Continued)*

(b)		
S. No.	**Organic Compounds**	
1.	Acetic acid, [(TMS)oxy],-TMS ester	
2.	Benzene, 1-ethyl-4-methyl	
3.	2-Butenedioic acid, bis(TMS) ester	
4.	Docosane	
5.	*n*-Pentadioic acid, bis(TMS) ester	
6.	Decanedioc acid	
7.	Ethanol, 2(octadecyloxy)-	
8.	Benzoic acid, 3,4,5-tris(TMS oxy), TMS ester	
9.	Hexadecanoic acid, TMS ester	
10.	Quercetin 7,3′,4′-trimethoxy	
11.	Octadecanoic acid, TMS ester	
12.	1*H*-Purin-6-amine,[(2-fluorophenyl)methyl]	
13.	Hexanedioic acid, dioctyl ester	
14.	Tetradecanoic acid, TMS ester	
15.	Hexadecanoic acid, 2,3-bis[(TMS)oxy] propyl ester	
16.	2-Monostearin TMS ether	
17.	Octadecanoic acid, 2,3-bis[(TMS)oxy] propyl ester	
18.	9,12-Octadecadienoic acid (Z,Z)-2,3-bis[(TMS)oxy] propyl ester	
19.	Glycocholic acid methyl ester TMS	

26.	Ergosten-3β-ol	
27.	Campesterol TMS	
28.	β-Sitosterol	
29.	Lanosterol	

(d)		
S. No.	**Organic Compounds**	
1.	Acetamide, 2,2,2-trifluoro-*N*-methyl-TMS	
2.	2-Butanol, *tert*-butyldimethylsilyl ether	
3.	Ethanedioic acid, bis(TMS) ester	
4.	Pyridine, 3-trimethylsiloxy	
5.	3-Hydroxy-6-methypyridine 1TMS	
6.	1,4-Dimethylpyrrolo(1,2-A) pyrazine	
7.	Benzene, 1,3-bis(1,1-dimethylethyl)	
8.	Docosane	
9.	2-Isoropyl-5-methyl-1-heptanol	
10.	Benzenepropanoic acid, TMS ester	
11.	Decanoic acid, TMS ester	
12.	Silane, trimethyl(undecycloxy)	
13.	Benzeneacetic acid, α,4-bis[(TMS)oxy], TMS ester	
14.	Emetan, 1′,2-didehydro-6′,7′,10,11-tetramethoxy	
15.	2,4-Imidazolidinedione, 1-[[(5-nitro-2-firnayl) methylene]amino	
16.	1-Hexadecanol, 2-methyl	
17.	Hexadecane	
18.	5-Hydroxy-2,2-dimethyl-5,6-bis(2-oxo-1-propyl)-1-cyclohexanone	
19.	Dodecenoic acid, TMS ester	
20.	2-Propenoic acid, oxybis(methyl-2,1-ethanediyl) ester	
21.	Benzoic acid, 2,5-bis(TMSoxy)-TMS ester	
22.	Dodecanoic acid	
23.	Heptacosane	
24.	1-Tetradecene, 2-decyl	
25.	1,2-Propanediol, 3-(octadecycloxy)-diacetate	
26.	β-Sitosterol trimethyl ether	
27.	2-Butanone, 4-[2-isopropyl-5-methyl	
28.	Hexadecanoic acid, methyl ester	
29.	2,5-Cyclohexadiene-1,4-dione, 2,6-bis(1,1-dimethylethyl)	
30.	Octadecanoic acid, methyl ester	
31.	17-Pentatriacontene	
32.	1,4-Ethano-1,2,3,4-tetrahydroanthracen-3-olbenzylidene	
33.	Pyrrolo [1,2-*a*]pyrazine-1,4-dione, hexahydro-3-(phenylmethyl)	
34.	Butyl 11-eicosenoate	
35.	Cholest-8(14)-en-3-one, (5α)	
36.	α-Homocholest-4*a*-en-3-one	
37.	Stigmasta-5,22-dien-3-ol(3β,22E)	

to reduce the volume of sludge generated. Sludge incineration can be characterized by technical and environmental problems and it could also face opposition from local communities. Dewatering can allow economical disposal or reuse of biosolids. Generally, chemical polymers are used for sludge settling and dewatering process in the wastewater treatment plants. Although landfilling of spent wash sludge could represent a sustainable solution, uncontrolled landfills and disposing of sludge to open areas has led to severe groundwater and soil pollution in many developing countries. To reach sustainable landfilling, mobile substances should be degraded, removed, or stabilized, landfill uncontrolled emissions be minimized, and the deposited materials should reach a final storage quality in equilibrium with the environment. Microorganisms can alter or break the sludge constituents. Recently, several filamentous fungi species are found to be useful. Some research works focused on the possibility of sludge to be recycled on tanning industry by using in biodiesel production, solidification, utilization in ceramics and bricks manufacture, or as a substitute of construction aggregates. Distillery sludge is the most common habitat that harbors unique types of bacterial species, which are capable of running widespread in situ bioremediation activities. The growing autochthonous bacterial species act at the primary level to loosen the interaction of organometallic bond through their enzymatic action, which makes metal available to plant. Generally, several indigenous plant species are excellent candidates for phytoremediation of distillery sludge.

3

Colorants of Distillery Waste and Their Properties

3.1 Introduction

The alcohol distillery industry is considered as one of the water-intensive, oldest, and largest industrial sectors in India, contributing around 14% to total industrial production of ethanol in India. In India, the Ministry of Environment, Forests, and Climate Change (MoEF&CC) have categorized the distilleries as a "red category" industrial sector, since it is considered a heavily polluting sector. In Indian distilleries, each liter of alcohol produced generates about 15 L of effluent or spent wash (Kumar and Chandra 2018a). The effluent generated from distilleries is highly colored and contains several refractory organic and inorganic pollutants (Figure 3.1a). The major color pollutants present in spent wash of ethanol fermentation processes are melanoidins and other Maillard reaction products (MRPs), phenolics, caramels, and hexose alkaline degradation products (HADPs) (Arimi et al. 2014). Among the color-causing pigments, melanoidins are the major dark brown color pigment of distillery effluent that are produced through non-enzymatic browning reactions such as the Maillard reaction (MR) as well as alkaline degradation reactions, or sugar degradation at medium temperature (>50 °C) and in basic pH medium (Echavarría et al. 2012; Chavan et al. 2013; Georgiou et al. 2016). As a result of the above compounds, distillery effluents are generally characterized by high biological oxygen demand (BOD), chemical oxygen demand (COD), and total suspended solids (TSS), total phosphorus (TP), ammonia nitrogen (NH_3-N), volatile suspended solids (VSS), suspended and dissolved solids, and soluble organic compounds; biodegradable organic materials, namely, carbohydrate, lignin, hemicellulose, dextrins, and organic acids were also commonly present in the distillery spent wash effluent and salts (sodium chloride) and have acidic (in spent wash) or alkaline (in biomethanated effluent) pH with high color index (Kaushik and Thakur 2009; Sharma et al. 2011; Kumar and Chopra 2012; Kumari et al. 2016; Shinde et al. 2020). In addition, distillery effluent also contains other refractory materials such as phenolic compounds, anthocyanins, tannins, and furfurans (e.g., hydroxyl methyl furfural) that can reach up to 10 g/L. Moreover, it also contains many toxic substances, such as toxic organic acids (butanedioic acid bis(TMS) ester; 2-hydroxysocaproic acid; benzenepropanoic acid, α-[(TMS)oxy], TMS ester; vanillylpropionic acid, bis(TMS)), and other recalcitrant organic pollutants (2-furancarboxylic acid, 5-[[(TMS)oxy] methyl], TMS ester; benzoic acid 3-methoxy-4-[(TMS)oxy], TMS ester; and tricarballylic acid 3TMS) (Chandra and Kumar 2017a). The high organic load of distillery effluent is mainly composed of melanoidins, which are produced through Maillard reaction (MR) between sugars and proteins and caramels from overheated sugars that are responsible for their color and odor (Watanabe et al. 1982; Tiwari et al. 2012; Taskin et al. 2016; Wilk and Krzywonos 2020). The high amount of melanoidins and various high- and low-molecular-weight Maillard reaction products (MRPs) and the presence of inorganic compounds like heavy metals make these effluents persistent organopollutants, which are extremely toxic and recalcitrant to natural biodegradation (Kumar and Chandra 2020a,b). Melanoidins and heavy metals also act as toxic, mutagenic, and carcinogenic agents, persist as environmental pollutants, and cross entire food chains providing biomagnification, such that organisms at higher trophic levels show higher levels of contamination compared to their prey (Chowdhary et al. 2018a; Kumar and Sharma 2019; Kumar et al. 2020a). In sugar processing, melanoidins are formed during purification and evaporation steps. Distillery industrial effluent is a heterogeneous mixture of androgenic-mutagenic and endocrine-disrupting compounds. Thus, the wastes produced from distilleries include fats, acids, and pigments, the majority of which are present in the effluent stream as well as sludge solids (Chowdhary et al. 2018a; Kumar

DOI: 10.1201/9781003029885-3

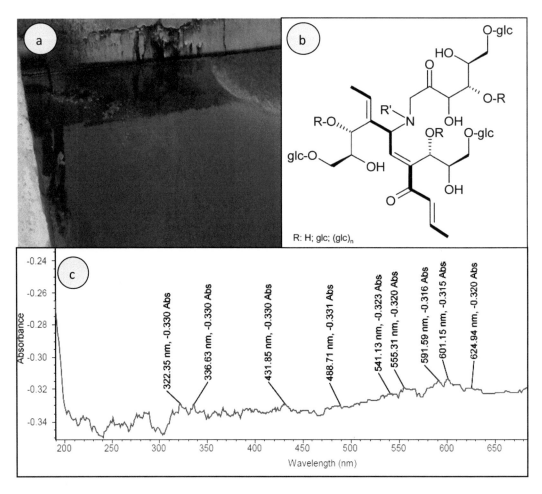

FIGURE 3.1 The complex nature of distillery effluent (a) Spent wash discharged after biomethanation (b) A typical structure of melanoidin molecule (c) UV-visible absorption spectrum of spent wash indicted various absorption peaks at different wavelength (λ) (Kumar et al. 2021).

and Chandra 2020b). Solid or liquid wastes derived from a series of pre- or postprocessing steps in the distillery industry could impose significant environmental impacts if not properly treated or managed. Therefore, the efficient and economical removal of coloring compounds from distillery effluents is an environmental challenge because of the difficulty of their removal from effluents (Kumar and Chandra 2020a; Chowdhary et al. 2020). Distillery effluent is not considered suitable for direct biological treatment processes due to its color and the presence of recalcitrant compounds. Effluents from the distillery industries are easily identifiable because of their intense colorization attributed to the presence of melanoidins. When these industrial effluents are discharged into natural water bodies (rivers and lakes), they cause serious perturbations as (i) reduction in sunlight penetration, which, in turn, decreases both photosynthetic activity and dissolved oxygen concentration (ii) persistence in the environment due to their bio-recalcitrance and toxicant effects due to the mutagenicity and/or carcinogenicity of their intermediate compounds, and (iii) alteration of the solubility of gases in the water. The impact of melanoidins-containing effluent in terms of toxicity, carcinogenicity, and genotoxicity has made the removal of color from the effluent more important than the removal of the soluble organic substances (Ramakritinan et al. 2005; Ayyasamy et al. 2008; Jain and Srivastava 2012; Malik et al. 2019b). In recent years, the interest of the distillery industry in wastewater treatments for reuse purposes has increased to be more efficient in the use of natural resources and to reduce its environmental impact.

Although conventional treatments such as biological and coagulation-flocculation processes are able to meet with the current regulations in terms of BOD and COD removal, they are not able for removal of color (Liang et al. 2009b; Liakos and Lazaridis 2014). Thus, potential treatment methods have to be investigated to remove color from distillery effluent, thereby preventing the serious environmental problem of decrease in dissolved oxygen concentration and photosynthetic activity in surface water resources due to draining of untreated waste. This chapter describes various color-causing compounds, including phenolic compounds, caramels, melanoidins, and melanin, that are responsible for the characteristic brown color of effluent discharges from distillery industries. This information may contribute to the design and management of cost-effective treatment technology.

3.2 Colorants in Distillery Effluent

Colorants that originate from the sugarcane plant include true pigments such as chlorophyll, anthocyanins, and flavonoids. Chlorophyll, being water insoluble, is easily removed, producing minor brown degradation products that are found in effluent. Anthocyanins, the red pigments, are also a minor component that is largely removed or destroyed in clarification, but a small amount is found in polymeric colorant in the spent wash. Flavonoids include a wide range of compounds that are in the yellow range, and which may react further to form highly colored polymeric colorant. Besides, an important group of plant-derived colorants and colorant precursors includes the many polyphenolic compounds, benzoic and cinnamic acid derivatives, which range from colorless to yellow, darkening with alkaline pH and able to produce highly pigmented iron complexes (Arimi et al. 2014; Zhang et al. 2017). Most process-derived sugar colorant is polymeric and has a high molecular weight. The phenolics are reactive and easily oxidized. Color formation involving enzymatic polymerization of polyphenolics in juice is estimated to be a major source of coloring material in both cane and beet processing (Godshall, 1996). Color in distillery effluent arises from a complex mixture of the following various organic chemicals:

i. Phenolic compounds, a number of organic compounds derived from the cane plant containing the highly reactive phenolic groups, are generated and extracted into the distillery effluent and sludge.
ii. Caramels, which are produced by thermal degradation and condensation reactions of sucrose.
iii. Melanoidins, a by-product of amino carbonyl sugar reaction called the MR.
iv. HADPs/alkaline degradation products of fructose, similar to caramels.
v. Melanin, reaction products of amino acids with phenolics as well as the very dark enzymatic oxidation products of phenolics. **Table 3.1** summarizes the typical colorant and their fate in environment discharges in distillery effluent.

3.2.1 Hexoses Alkaline Degradation Products (HADPs)

HADPs together with melanoidins are responsible for up to 80% of color in sugar beet juices. Monosaccharides in aqueous alkaline solutions undergo both reversible and irreversible transformations (Guimarães et al., 1999). The reversible reactions, ionization, mutarotation, enolization, and isomerization, result in the formation of the enediol anions that are generally considered common intermediates in isomerization reactions of monosaccharides. Enediol species are considered to be intermediates in the isomerization of monosaccharides, as well as starting intermediates in the alkaline degradation reactions.

3.2.2 Polyphenols

Polyphenols are natural antioxidants containing phenol group(s) in their structure and have attracted lots of interest in current scenario due to their special properties, including antioxidant, antimicrobial, and anticarcinogenic activities. Polyphenols are categorized into three major classes: phenolic acid,

TABLE 3.1

The Major Distillery Effluent's Colorant, Their Properties, and Impact on the Environment

S. No.	Colorant	Description	Environmental Impact
1.	Phenolics	The major phenolic acids present in distillery effluent, including cinnamic acid and its derivatives, such as caffeic acid, *p*-coumaric acid, ferulic acid, and chlorogenic acid, and benzoic acid and its derivatives	Phenolics give a high antimicrobial antioxidant activity to effluent, thus slowing down the aerobic and anaerobic treatment processes. It is established that phenolic compounds in wastewater are major contributors to toxicity, limiting its microbial degradability
2.	Caramels	They are colloidal compounds with a tendency to remain preferentially on the crystal surface, thus affecting the quality of white sugar. Heating concentrated sucrose syrups at temperatures above 210 °C forms caramels	At present, there is no evidence that caramel has a significant antimicrobial effect. The colloidal nature of caramels makes them resistant to decomposition and toxic to microflora
3.	Melanoidins and MR products	Melanoidins are formed during the reaction between amino compounds and carbohydrates. Melanoidins are generally regarded to be heterogeneous, acidic, high-molecular-weight (5–40 kDa), recalcitrant polymers with chemical properties similar to humic substances (i.e., humic and fulvic acids). It is composed of highly dispersed colloids, which are negatively charged due to the dissociation of carboxylic acids and phenolic groups	Melanoidin is antioxidant color pigments and primary treatments involving biological treatments were found ineffective to degrade the color compounds at high concentrations. Melanoidins have shown antimicrobial activity against different microbial species
4.	Hexose alkaline degradation products	Hexose alkaline degradation products together with the melanoidins are responsible for up to 80% of color in sugar beet juices	—

Note: Amino acids and reducing sugars are not colorants, but their presence and the Maillard reactions (MRs) they undergo contribute to color formation in processing.

flavonoids, and tannins. The major phenolic acids have been detected in distillery effluent, including cinnamic acid and its derivatives (caffeic acid, *p*-coumaric acid, ferulic acid, and chlorogenic acid) and benzoic acid and its derivatives (e.g., gentisic acid, gallic acid), which give a high antimicrobial activity to this effluent, thus slowing down the aerobic and anaerobic treatment processes. Several studies have highlighted the antioxidant activity of both phenolic compounds and MRPs.

3.2.3 Caramels

Caramel is a brownish-black viscous liquid or a hygroscopic powder with a high molecular weight (>10 kDa), and caramels from sucrose have been used commercially as food colorants for decades and are so far still the most popular in the food industry. Caramels, thermal degradation products of sugars, are formed in a complex series of reactions involving the dehydration of carbohydrates (predominantly monosaccharides) to reactive species, which then polymerize together. In the process of sugar production, the caramelization reaction occurs mainly in the crystallizer when sucrose syrups are subjected to high temperatures (>210 °C) and the process can be catalyzed by acid and base. The generation of color in caramelization requires that sugars, normally monosaccharide structures, should first undergo intramolecular rearrangements. Depending on the time and temperature, yellow or brown solutions are obtained. In the sugar degradation reactions, osuloses are formed, which are considered to be intermediates of caramelization.

3.2.4 Melanoidins and Maillard Reaction Products

Melanoidins are intense dark brown, complex, organic polymers present in food items, such as coffee, honey, and beer, and biological material and have significant effects on the quality of food since colors and flavors are important food attributes and a key factor in consumer's acceptance (Hayase et al. 1984; Echavarría et al. 2012). Moreover, melanoidins are a major nitrogenous color imparting organic compounds produced in molasses-based ethanol distilleries that make a significant component (nearly 2%) of effluent (spent wash). The empirical formula of melanoidin is $C_{17-18}H_{26-27}O_{10}N$ and their structure is illustrated in Figure 3.1b. Melanoidins, in distillery effluent, formed through the MR or non-enzymatic browning reaction occur between amino acids and carbohydrates at temperatures above 50 °C and pH 4–7 (Hayase 2000; Arimi et al. 2014). The MR encompasses a network of various reactions between reducing sugars and compounds with a free amino group forming a variety of products, which can be classified as initial stage products, intermediate stage products, and advanced stage products. The end products of MR are called melanoidins and are generally defined as high-molecular-weight (usually 5–40 kDa), heterogeneous compounds (Moreira et al. 2012; Langner and Rzeski 2014). Melanoidins are difficult to characterize due to their varying sizes and types of reducing sugars and amino acids involved in their formation. The negative charge of melanoidins is due to the dissociation of carboxylic and phenolic groups. It has been reported that due to their net negative charge, various metallic ions such as Cu^{2+}, Cr^{6+}, Fe^{3+}, Zn^{2+}, Pb^{2+}, etc. bind with melanoidins to form an organometallic complex (Hatano et al. 2016; Migo et al. 1997). Consequently, the high metallic ions-binding tendency of melanoidins also enhances the vulnerability of organometallic complex toward its toxicity in the ecosystem (Chandra et al. 2008). In sugar processing, melanoidins are formed during purification and evaporation steps. MRPs are a particularly complex mix of various compounds of different molecular weights. They include not only aldehydes, ketones, dicarbonyls, acryl amides, and heterocyclic amines, all of which contribute to flavor, but also melanoidins and advanced glycation end products, which are polymeric products formed at the advanced steps of MR (Echavarría et al. 2012; Langner and Rzeski 2014). MRPs also have antioxidant properties, as shown for coffee, vinegar, processed food such as beer, pasta, or tomato puree, and model systems (Moreira et al. 2012; Caderby et al. 2013; Kaushik et al. 2018). The presence of melanoidin leads to the dark brown color of the distillery effluent (Kumar et al. 1998). Due to the recalcitrant antioxidant nature of melanoidins, it inhibits microbial growth used in biological treatment processes of distillery effluent (Caderby et al. 2013; Kaushik et al. 2018). Consequently, distillery effluent is a major source of aquatic and terrestrial pollution in the environment. The structure of melanoidins is not yet fully understood, but many research efforts have been done to determine their structure and chemical properties. The elemental and chemical structure of melanoidins depends heavily on the nature and molecular concentration of reactant and reaction conditions such as pH, heating time temperature, and solvent. Cammerer et al. (2002) have proposed the fundamental structure of melanoidins pigment formed from Amadori reaction products and 3-deoxyhexosuloses, as shown in Figure 3.1b. The discharged distillery effluent contains a mixture of Maillard products (i.e., initial, intermediate, and advanced stages with variable molecular weight) (Kumar and Chandra 2018a). Melanoidins have physiological characteristics of antimicrobial, antioxidant, and antihypertensive activities, antioxidative and radical-scavenging activities, metal chelating, antibacterial, antioxidant, and antiallergenic (Moreira et al. 2012; Langner and Rzeski 2014; Kaushik et al. 2018). However, when melanoidins are present in industrial effluents such as sugar refineries and fermentation industries, they are hazardous for the ecosystems (e.g., cytotoxic and the antimicrobial factors, severe soil, and water pollutants) (Arimi et al. 2014; Kumar and Sharma 2019; Kumar and Chandra 2020a). Consequentially, melanoidin effluent pretreatment is demanded to avoid contamination around the outflow of these refineries.

3.3 Properties of Melanoidins

   i. Melanoidins are dark brown recalcitrant polymeric macromolecules of distillery effluents with high BOD and high COD.

ii. Although melanoidins are chemically diverse, many studies generally agreed that melanoidins are negatively charged, brown-colored, high-molecular-weight, nitrogen-containing compounds analyzed in both real foods and model systems.

iii. Melanoidins are generated as the end products of MR and formed primarily by interactions between amino acids and carbohydrates. MR occurs when a carbohydrate reacts with an amino acid or an amine at a temperature higher than 50 °C and pH 4 7.

iv. Melanoidins have antioxidant, anti-hypertensive, and antimicrobial activities.

v. Melanoidins in wastewater prevent sunlight penetration and reduce both photosynthetic activity and the dissolved oxygen level in the aquatic ecosystem.

3.4 Maillard Reaction

The MR has been named after the French renowned chemist Louis Maillard who first described this reactions in 1912; however, it was only in 1953 that the first coherent scheme was put forward by John Hodge. John Hodge published his consolidated scheme that summarized the chemical reactions that were understood to comprise the MR at that time (Wang et al. 2011). The Hodge scheme remains widely used today (Figure 3.2). The MR is a complex series of reactions and initiated by a condensation reaction between the carbonyl group of the aldose and the free amino group of an amino acid to give an N-substituted aldosylamine. The condensation product rapidly loses water molecules as a product and is converted into a Schiff base. The resulting Schiff base cyclizes in the case of pentoses and hexoses to the corresponding glycosylamine, which then undergoes an Amadori rearrangement. If the aldose and amino acid glycine, then the Amadori product is 1-amino-1-deoxy-2-fructose (monofructose glycine), with fructose the reaction is quite similar but the rearrangement is termed the Heyns rearrangement and is generally shown as giving substituted 2-amino-2-deoxyldoses. The Amadori rearrangement is considered to be the key step in the formation of major intermediates for the browning reaction. Aminal doses are not very stable and readily react forming the Amadori compound. The second stage in the MR is usually depicted as starting with the decomposition of Amadori and Heyns adducts to form deoxydicarbonyl sugars. Amadori product and its dicarbonyl derivatives can undergo concurrently retro-aldol reactions producing more reactive C_2-C_5 sugar fragments, like diketones, glyceraldehydes, and hydroxyacetone derivatives. This chemical reaction is called Strecker degradation, and it is characterized by the production of carbon dioxide. Strecker degradation products especially during cooking, and many of the heterocyclic compounds that cause flavor and aroma, are formed at this stage. Finally, the condensation of some of the products formed in this step is produced either among them or with amino compounds to form brown pigments and polymers (Echavarría et al. 2012). The major steps involved in melanoidins formation from sugar and amino acid reactions are illustrated in Figure 3.2.

Despite efforts in recent years, the chemical structure of melanoidins is still not completely understood, but it is assumed that it does not have a definite structure as its elemental composition and chemical structures largely depend on the nature and molar concentration of parent reacting compounds and reaction conditions like pH, temperature, heating time, and solvent system used. Melanoidins are highly resistant to degradation due to their complex chemical structures. Melanoidins are characterized by the presence of one or more C=C and C≡N bonds with phenolic rings. The absorption of electromagnetic radiations in the ultraviolet and visible regions by a molecule causes the electronic excitation and an electron moves to higher electronic energy level from a lower. A covalently unsaturated group responsible for absorption in the UV or visible region is known as a chromophore. For example, C=C, C≡C, C=O, C≡N, N=N, NO_2 etc. If a compound absorbs light in the visible region (400–800 nm), only then it appears colored. Thus, a chromophore may or may not impart color to a compound, depending on whether the chromophore absorbs radiation in the visible or UV region. Chromophores like C=C or C≡C having π electrons undergo π → π* transitions and those having both π and non-bonding electrons, e.g., C=O, C≡N or N=N, undergo π → π*, n → π* and n → σ* transitions. Since the wavelength and intensity of absorption depend on several factors, there are no set rules for the identification of a chromophore. Melanoidins possess characteristic chromophoric groups that usually contain pyrroles, imidazoles, and

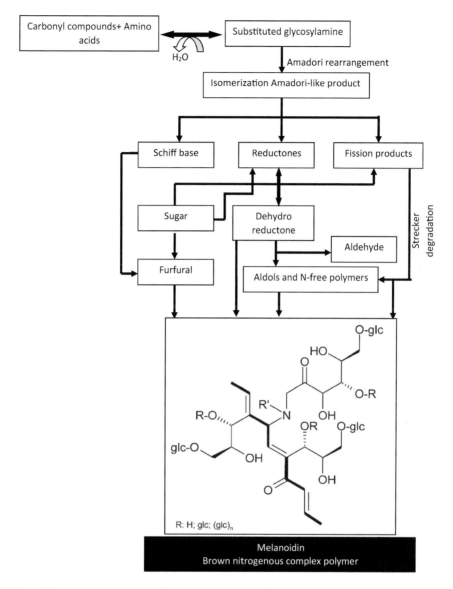

FIGURE 3.2 Schematic representation of major steps occurs in melanoidin formation from amino acids and carbohydrates.

their nitrogen-containing derivatives, which are responsible to impart colors to compounds. They exhibit the ability to absorb light at 475 nm, which enables them to be quantified in, for example, coffee brews, other foodstuffs, model systems, or distillery effluent. The color intensity of melanoidin becomes higher with the degree of polymerization. Therefore, melanoidin shows independent absorbance maxima (λ_{max}) at different wavelengths of light, as shown in Figure 3.1c. However, melanoidins generally show variable absorption range in the UV region, which makes it more difficult to understand the mechanism of melanoidins degradation and decolorization and characterization of its metabolic products. Most of decolorization and degradation of melanoidins in model system or in discharged effluent is reported at λ_{max} of 475 nm using a UV-visible spectrophotometer because melanoidins absorb light at wavelengths as high as 470 nm and are predominantly responsible for the characteristic brown color of distillery effluent. As shown in Figure 3.1c, variable absorption spectrum of spent wash at different wavelengths indicated the presence of a mixture of different molecular weight MRPs in distillery effluent. This property makes it easier to monitor melanoidin removal over time. Decrease or change in absorbance clearly means that the melanoidins are being removed or transformed and can easily be measured using a simple colorimeter or UV-visible spectrophotometer. Because melanoidins cannot be directly analyzed due to the uncertainty of their structures, they are usually quantified by difference, subtracting the total percentage of known compounds from 100 percentage products, the melanoidins. The % decolorization was calculated using the below equation:

$$\text{Decolorization } (\%) = A_i - A_f \ / \ A_i \times 100$$

where

A_i is the initial absorbance and A_f is the final absorbance of the medium after decolorization at the λ_{max} in nm.

4

Environmental Impacts and Health Hazards of Distillery Waste

4.1 Introduction

High volumes of highly colored effluent produced during ethanol production in distillery industry is considered to be one of the most polluted industrial effluents consisting of high concentration of organic compounds and other toxicants (Godshall 1996; Ziaei-Rad et al. 2020). Since the distillery effluent contains highly stable suspended color pigments (i.e., melanoidins) that cannot be separated with conventional biological and physicochemical treatment methods, there is always a lookout for advanced treatment methods. However, most of the distilleries dispose of their partially treated or untreated effluent into water bodies causing environmental threats to organisms (Singh et al. 2010; Padoley et al., 2012; Malik et al. 2019). This increasing toxicity of discharged effluents affects the human beings in several ways making melanoidins contamination both environmental and public health issues (Mahar et al. 2013, Chauhan and Rai 2010). Thus, disposing of distillery effluent into the environment without adequate treatment is hazardous to the ecosystem and has a high pollution potential (Singh and Nigam 1995; Rodriguez-caballero et al. 2012). Some essential metals necessary for plant metabolism as enzyme activators or cofactors, e.g., Fe, Cu, Mn, Zn, and Ni, are also present in distillery effluent (Chandra et al. 2018c; Kumar and Chandra 2020a,b). Besides, the presence of minerals and metallic ions, distillery effluents also contain significant amounts of recalcitrant organic compounds. Inclusion of heavy metals and organic compounds in receiving aquatic reservoirs alters the vital parameters of water bodies by influencing the levels of chemical oxygen demand (COD), biological oxygen demand (BOD), total organic carbon (TOC), total dissolved solids (TDS), total suspended solids (TSS), pH, suspended solids, sulfide, color, and ammonical nitrogen that enhance the hardness of the water (Yadav and Chandra, 2011; Alves et al., 2015). It can alter oxygen levels and pH can impede the penetration of light in the water causing disruption of the aquatic ecosystem and is potentially toxic and mutagenic to aquatic flora and fauna. Some investigators report that furfurals, melanoidins, and their metabolites are toxic, carcinogenic, and mutagenic, which inhibits the growth of bacteria, protozoa, algae, plants, and different animals (Chandra and Kumar 2017a, Chowdhary et al. 2020). The potential adverse effects on humans and the entire ecosystem due to the direct disposal of textile effluent into the aquatic environment without proper treatment have been widely reported. Effluent stored in unlined lagoons in distillery percolate into the groundwater table and bore wells discharge light tea-colored sharbat in place of colorless drinking water (Mahar et al. 2013). Due to frequent occurrences of such incidents and more due to depleting reserves of clean drinking water, people have become more conscious about distillery waste disposal. One can see very frequent news items appearing in newspapers about health hazard and environmental degradation due to untreated or partially treated distillery effluents. The release of melanoidins-containing effluent from both small- and large-scale fermentation and distillery industry into the environment can be an ecotoxic hazard and can affect man through the potential danger of bioaccumulation by transport through the food chain (Ayyasamy et al. 2008; Chauhan and Rai 2010; Asano et al. 2014). These colored industrial effluents are often contaminated with harmful or poisonous chemical pollutants when withdrawn on the land, affecting the germination rate of several plants and thereby decreasing the soil fertility. It inhibits germination of seeds and depletes vegetation by decreasing the soil alkalinity, salinity, and manganese availability. It has also been reported that the distillery effluent is toxic to plants (Sinha et al. 2014). Thus,

DOI: 10.1201/9781003029885-4

the appropriate dilution of the textile effluent is necessary before it is used and may help to reduce the toxic effects of distillery effluent and establishes the correct proportion of nutrient fertilization in soil (Kumari et al. 2016). Keeping the widespread impacts of distillery effluent in view, their treatment is a topic of interest in recent research. Therefore, addressing the issue of these environmental contaminants has become very important and needs attention from environmentalists. The following sections of this chapter discuss the impacts of distillery effluent on soil and ground water quality, ecotoxicological hazards of spent wash, as well as the positive and negative impact of untreated or partly treated distillery effluent into the environment

4.2 Irrigation of Soil and Impacts on Groundwater Quality

During the past two decades, the reuse of treated or untreated effluent in irrigation has been quite common in Europe, the United States, Mexico, Australia, China, India, and the Near East and, to a lesser extent, in Chile, Peru, Argentina, Sudan, and South Africa. In developing countries, industrial effluent has always been a low-cost option for irrigated agriculture to the farmers, especially in water-scarce regions in water-starved arid and semiarid parts of tropical countries (Kaushik et al. 2005; Chauhan and Rai 2010; Kumari et al. 2016). Irrigation with treated or untreated industrial effluent is a relatively recent practice since it is seen (a) as a low-cost option for effluent disposal, (b) as a reliable and cheap source for irrigated agriculture, (c) as a way of keeping surface water bodies less polluted, and (d) as an important economic resource for agriculture due to its nutritive value (Arora et al. 1992; Asano et al. 2014; Jain and Srivastava 2012). The reuse of wastewater for purposes such as ferti-irrigation saves the amount of water that needs to be extracted from environmental water sources. Most of the distillery units in India have opted distillery effluent for agricultural irrigation. Molasses-based distilleries generate large quantities of effluent, which is being used for irrigation in many countries including India. In ferti-irrigation, the yield of wheat (*Tritium aestivum*) increased by 33% as compared to the control using di-ammonium phosphate (DAP) and urea (Kumari et al. 2009). Kumari et al. (2012) suggested that the diluted distillery effluent is capable of replacing the application of chemical fertilizer when used under controlled conditions without any adverse effect on the soil and groundwater quality. The effluent is rich in organic and inorganic ions, which may leach down and pollute the groundwater. A study has been conducted by Jain et al. (2005) to assess the impact of long-term irrigation of post-methanation distillery effluent on groundwater quality. This study indicated that long-term indiscriminate use of post-methanation distillery effluent leads to significant leaching of organic and inorganic ions and pose a serious threat to the groundwater quality if applied without proper monitoring. Similarly, Chauhan and Rai (2010) showed that the irrigation with molasses-based distillery effluent impaired the groundwater quality of Gajraula region, especially of agricultural zone, making it unsuitable for drinking purposes. Mahar et al. (2013) examined the effect of heavy metals such as Cd, Co, Cr, Cu, Mn, Ni, Pb, Zn, and As concentration in groundwater quality sampled from the vicinity of distillery spent wash evaporation ponds. They concluded that the distillery spent wash is the cause for the change in heavy metal contents of the study area. The concentration of metals like Mn, Cd, Cu, Fe, Ni, and Zn in groundwater samples had indicated higher values than their respective permissible limits, while heavy metals concentration in reference samples was found within limits of world health organization (WHO) for drinking water. Diluted spent wash could be used for irrigation purposes without adversely affecting soil fertility, seed germination, and crop productivity. Ayyasamy et al. (2008) conducted a pot experiment to study the effects of different concentrations (20%, 40%, 60%, 80%, and 100%) of sugar factory effluent on seed germination, seedling growth, and biochemical characteristics of green gram and maize. A similar study was also carried out using the aquatic plants, water hyacinth, and water lettuce. The higher effluent concentrations (>60%) were found to affect plant growth, but diluted effluent (up to 60%) favored seedling growth. This study concluded that physicochemical parameters, such as BOD, chloride, alkalinity, hardness, calcium, magnesium, sulfate, and phosphate were relatively higher in the sugar factory effluent and severely affected the plant growth. There was a gradual decrease in the shoot length and free amino acid, protein, and total chlorophyll contents in both terrestrial and aquatic plants when irrigated with various effluent concentrations compared to the control. The untreated effluent could lead to soil pollution, deterioration, and low productivity. Both the terrestrial and aquatic environments were affected, which could be averted by proper treatment of the effluents using suitable conventional methods. The diluted spent wash irrigation improved the physicochemical properties of the soil and further increased soil microflora. Many farmers in the vicinity of sugar factories apply spent wash and its products as manure. However, as the spent wash contains a considerable quantity of salt, its indiscriminate use for a long period has a risk of deteriorating the physicochemical properties of the soil. Nutrient composition of crude and digested spent wash and its impact on sugarcane growth and biochemical attributes have been studied by Jain and Srivastava (2012). Authors indicated a stimulatory effect on root and shoot growth at a low rate of crude spent wash (5 mL/kg soil) and inhibitory effect of higher dose (100 mL/kg soil) of both crude and digested spent wash; therefore, judicious application of spent wash will improve crop productivity and alleviate environmental pollution problems. Tripathi et al. (2011) indicated the effect of the application of various dosages of distillery effluent irrigation on soil physicochemical, cellulase, and urease activities in a tropical agricultural field. This study clearly indicated that discharge of effluents from distillery has altered the physicochemical properties and enhanced the cellulase and urease activity of the soil, but it declined with time. The enzyme activity was improved by up to 50% and later decreased. Also, a suitable treatment of distillery effluent is essential to remove heavy metals and other toxic organic compounds. Similarly, Narain et al. (2012) observed that the higher concentrations of distillery effluent altered the physicochemical and structural properties of the soil. The appearance of inorganic nitrates, inorganic carbonate, aliphatic hydrocarbon, and aliphatic primary amides at various concentrations of distillery effluent showed a change in functional groups and infrared spectra, probably due to the differential breakdown of parent component present in the distillery effluent. As the concentration of the effluent increases, the transmittance decreases, which shows that the soil structure is disturbed at higher concentration because of the presence of heavy metals or some other contamination present in the effluent. Moreover, numerous reports indicate that distillery effluents have toxic effects on the germination rates and biomass of several plants species, which have important ecological functions, such as providing habitat for wildlife, protecting soil from erosion, and providing the organic matter that is so significant to soil fertility. Kumar and Chopra (2012) studied the fertigation effect of distillery effluent concentrations such as 5%, 10%, 25%, 50%, 75%, and 100% on agronomical practices of *Trigonella foenum-graecum* L. (Fenugreek). It was observed that irrigation improved soil nutrient status. The effluent has potentiality for its use as agro-based biofertigant in the form of plant nutrients needed by *T. foenum-graecum* crop plant. Therefore, it can be used as agro-based biofertigant after its appropriate dilution for irrigation purposes for the maximum yield of this crop and to maintain soil health. Sugarcane stillage is frequently applied to field crops (mainly to sugarcane plantations) diluted with irrigation water or sugar cane wash water. However, the low pH, electric conductivity, and chemical elements present in stillage may cause changes in the physicochemical properties of soils, rivers, and lakes with frequent discharges over a long period. In addition, it might also have adverse effects on agricultural soils and biota in general (Christofoletti et al. 2013). Taking into consideration the environmental risk of stillage and the large generated volumes (10–15 L/L ethanol), several alternatives have been developed for its treatment before being applied to field crops or discharged into bodies of water, such as anaerobic digestion (Bhoite and Vaidya, 2018a) and phytofiltration using constructed wetlands (Sánchez-Galván and Bolaños-Santiago 2018). Figure 4.1 presents a simplified view of the environmental pollution caused by indiscriminate disposal of color effluent.

4.3 Ecotoxicological Hazards of Spent Wash

The fast-growing distilleries are one of the world's most pollution-generating industries, and management of its toxic large quantities of colored effluents has become a global issue. The presence of androgenic-mutagenic compounds in distillery effluent and their discharges in ecosystems present serious environmental and health concerns as a result of the toxicity of the compounds themselves and their transformation into toxic, mutagenic, and carcinogenic compounds (Figure 4.2). The impact of distillery effluent and distillery soil leachate on vertebrates species, such as freshwater Teleost, *Channa punctatus* (Kumar and Gopal 2001), freshwater fish, *Cyprinus carpio* (Ramakritinan et al. 2005) and male mice,

FIGURE 4.1 Environmental pollution due to discharges of untreated or partly treated distillery effluent (a) Color effluent generated during bioethanol production (b) Aquatic pollution (c-e) Soil and groundwater pollution due to leaching of disposed effluent.

Mus Musculus L. (Sharma et al. 2011), and invertebrate species, such as *Eisenia andrei*, *Enchytraeus crypticus*, *Hypoaspis aculeifer*, and *Folsomia candida*, were assessed through ecotoxicological assays. Kumar and Gopal (2001) examined the impact of distillery effluent on physiological consequences on *C. punctatus*. They stated that distillery effluent affects the tissue protein, lactic acid in tissues, blood glucose and glycogen level, and various blood parameters of *C. punctatus* exposed to different concentration for 96 h. Similarly, Ramakritinan et al. (2005) studied the impact of distillery effluent on carbohydrate metabolism of *C. carpio* at different days during exposure (7, 14, and 21 days) in the ambient temperature of 28 ± 1 °C. Oxygen, total carbohydrate, glycogen content, and succinate dehydrogenase (SDH) activity in muscle, liver, and brain tissues of *C. carpio* exposed to different sublethal concentrations decreased gradually and significantly. However, serum glucose and lactic acid content showed an increasing trend with increase in effluent concentration and time of exposure. Unlike SDH, lactic acid

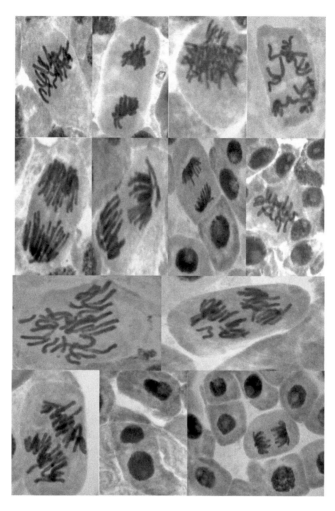

FIGURE 4.2 Different chromosome aberration observed in meristematic cells of *Allium cepa* ($2n = 16$) treated with distillery effluent.

dehydrogenase (LDH) activity of muscle, liver, and brain tissues showed an increasing trend and the enhancement of enzyme activity was more in liver tissue (71.3%). Sharma et al. (2011) evaluated the toxic effect of different concentrations (5–20%) of distillery soil leachate on reproductive functions of *Swiss albino* male mice. The accessory reproductive organ (epididymis) is an important toxicological target to pollutants present in distillery soil leachate. They observed that body and epididymis weight decreased in all concentration groups of distillery soil leachate treated animals. Moreover, the total protein content of epididymis (36–75%), sperm count (78–87%), and sperm motility (20–33%) decreased significantly in experimental animals. Furthermore, histological changes were observed in epididymis. In addition, the morphological abnormalities were also seen in sperms. The highest concentrations of vinasses caused avoidance behavior in earthworms and collembolans, and reduced the number of earthworm juveniles in natural soils (Alves et al. 2015). The high colors render the water unfit for use at the downstream of the disposal point and may hinder light penetration, thereby affecting aquatic life and continuously threatening the biodiversity. Asano et al. (2014) assessed the effects of 37 microbes on plant-available inorganic nutrients and a phytohormone and their interaction in the bioconversion of rice-derived distillery effluents into liquid fertilizer. Inoculation of several *Aspergillus* and *Bacillus* strains resulted in accumulation of a large quantity of ammonium (NH_4-N) in the effluent. However, a decrease in the liquid phase during *Aspergillus* incubation suggested the requirement for additional treatment of the solid residue, whereas the growth of *Bacillus subtilis* inhibited by the acidic conditions in the raw distillery effluent suggests the requirement for pH adjustment prior to incubation. Interestingly, *Aspergillus caelatus*, *A. oryzae*, and *A. tamarii* yielded greater increases in nitrate (NO_3^-) concentrations (30–39 mg/L). Indole-3-acetic acid (IAA) levels increased significantly in pH-adjusted effluents inoculated with *Wickerhamomyces*. Colorimetric and gas chromatography-mass spectrometry (GC-MS) analyses revealed that *Wickerhamomyces* strains generated 7–26 mg/L of IAA. Kumari et al. (2019) studied the plant-growth-promoting (PGP) traits (phosphate solubilization, siderophore, indolic compounds, and ammonia production) and taxonomic composition of bacterial communities in agricultural soil irrigated with distillery effluent in conjugation with irrigation water, using cultivation-dependent and cultivation-independent methods. The cultivation-dependent and cultivation-independent methods provided a holistic picture of the bacterial community composition in distillery-effluent-irrigated soil. Diverse bacterial taxa were found in both culturable bacterial community and 16S rRNA gene clone library, which belonged to bacterial phyla Proteobacteria (Alpha, Beta, and Gamma subdivisions), Firmicutes, Actinobacteria, Acidobacteria, Bacteroidetes, and Gemmatimonadetes. Distillery effluent irrigation did not seem to have an adverse effect on PGP traits of the culturable bacterial community; therefore, using effluent in conjugation with irrigation water could be a viable water reuse method in regions facing water scarcity. Most of the bacterial isolates obtained from distillery-effluent-irrigated soil were found to display PGP traits. Chandra and Kumar (2017a) reported the toxic effects of spent wash at different concentrations on seedling growth of *Phaseolus mungo* L. and *Triticum aestivum*. They found that a high concentration of spent wash affects the seedling growth of *P. mungo* L. and *T. aestivum*. Chandra and Kumar (2017b) also reported the presence of androgenic-mutagenic compounds and potential autochthonous bacterial communities during in situ bioremediation of anaerobically digested distillery sludge and also tested the toxicity of in situ degraded sludge leachate by using *Allium cepa* L. root meristematic cell. They showed a reduction of toxicity in degraded samples of sludge and leachate, confirming the role of autochthonous bacterial communities in the bioremediation of distillery waste in situ. Cassman et al. (2018) characterized the microbial assemblage of vinasse to determine the gene potential of vinasse microbes for contributing to negative environmental effects during ferti-irrigation and/or to the obstruction of bioethanol fermentation. The vinasse microbial assemblage was characterized by low alpha diversity with 5–15 species across the six vinasses. The core genus was *Lactobacillus*. The top six bacterial genera across the samples were *Lactobacillus*, *Megasphaera*, and *Mitsuokella* (Phylum Firmicutes, 35–97% of samplereads); *Arcobacter* and *Alcaligenes* (Phylum Proteobacteria, 0–40%); *Dysgonomonas* (Phylum Bacteroidetes, 0–53%); and Bifidobacterium (Phylum Actinobacteria, 0–18%). Potential genes for denitrification but not nitrification were identified in the vinasse metagenomes, with putative nirK and nosZ genes the most represented. Binning resulted in 38 large bins with 36–99.3% completeness and 5 small mobile element bins. Of the large bins, 53% could be classified at the phylum level as Firmicutes, 15% as Proteobacteria, 13% as unknown phyla, 13% as Bacteroidetes, and 6% as Actinobacteria. Chowdhary et al. (2018b) investigated the effects of potentially toxic elements on biochemical parameters in wheat (*T. aestivum* L.) and mustard (*Brassica juncea* L.) plants growing at distillery and tannery wastewater contaminated sites. The decline in plant protein, carbohydrates, and chlorophyll is a clear indication of the toxic effects of distillery effluent on plants. This study concluded that industrial wastewaters are the primary sources of metal accumulation in agricultural crops and thus it should not be discharged into the environment before its proper treatment.

4.4 Positive Impact of PMDE

Post-methanation effluent (PME) generated through bio-methanation of spent wash, a foul-smelling, dark-colored by-product of distillery industries, is applied to arable land in some areas near the vicinity of the distillery industries as an amendment. The PME contains considerable amount of organic matter and salt besides its high plant-nutrient content. Impact of long-term effluent irrigation in the field (10 years) and short-term effluent irrigation using different doses of PME in the laboratory (30 days) was studied in combination with three bioamendments, i.e., farmyard manure, brassica residues, and rice husk (Kaushik et al. 2005). Long-term application of PME proved useful in significantly increasing

TOC, TKN, K, P, and soil enzymatic activities in the soil but tended to build up a harmful concentration of Na that could be chelated by bioamendments. In short-term studies, application of 50% PME along with bioamendments proved to be the most useful in improving the properties of sodic soil and also favored successful germination and improved seedling growth of pearl millet. Hati et al. (2004) indicated that application of PME to the agricultural field, as an amendment, might be a viable option for the safe disposal of this industrial waste with concomitant improvement in yield and physical properties of the soil. However, the level of PME application should be within the prescribed limit to avoid probable development of soil salinity in the PME-treated field due to its high salt load. Several negative effects of organic and inorganic pollutants on different living organisms due to melanoidins and distillery effluent exposure have been evidenced. Narain et al. (2012) studied the Impact of posttreated effluent distillery effluent on germination and seedling growth of *Pisum sativum* L. The physicochemical characteristics of the effluent indicate that it is alkaline and rich in chlorides and TDS. Effluent color is dark brown and has a pungent smell. Distillery effluent did not show any inhibitory effect on seed germination, vigor index, root length, shoot length, and dry weight at a lower concentration (25%). The integrated effect of biomethanated distillery effluent (BMDE), bio-compost, and FYM as a source of plant nutrients and their effect on soil properties, nutrient uptake, juice quality, and sugarcane yield were investigated by Sinha et al. (2014). They stated that integrated use of BMDE along with inorganic potassium was superior to other treatments improving cane yield. The quality of juice, namely, sucrose and purity, remain unaffected. However, commercial cane sugar significantly increased in all the treatments over control: the application of different doses of BMDE and bio-compost and K_2O content of soil over control (recommended dose). They also enriched the soil organic matter that improved the physical properties of soil, especially the water transmission characteristic of soil. Hydraulic conductivity, infiltration, and soil aggregation increased significantly due to application of bio-compost and BMDE. The soil aggregation increased and bulk density decreased with increasing levels of BMDE. The K content of postharvest soil increased substantially in BMDE and bio-compost-treated plots. The application of BMDE and bio-compost brings remarkable changes in the properties of soil and thus enhance the fertility of soil and productivity of sugarcane significantly. The bio-composting not only solves the disposal problems of effluent but also helps in saving the cost of chemical fertilizers for sustainable sugarcane production. The result also indicated that integrated use of BMDE along with inorganic potassium was superior in terms of improving cane yield. It also enriches the soil in terms of organic matter that improved the physical properties of soil, especially the water transmission characteristic of soil. Onetime controlled land application of BMDE as liquid manure saved 100% potassium fertilizer and also enriched soil fertility for obtaining profitable cane yield in calcareous soils. A promising approach to recycle the residual distillery waste as a potential liquid fertilizer has been reported by Kumari et al. (2016). Field studies were conducted on *Brassica compestris* to assess the potential of the diluted post-methanated distillery effluent (PMDE). There is a significant increase in the height, weight, length of the root, and number of root hairs, area of the leaf, number of branches per plant, number of siliquae per plant, and number of seeds per siliquae, resulting in greater seed weight and higher hydrophilic colloids, indicating higher oil yield when treated with the diluted PMDE. Hence, the application of nutrient-rich PMDE to the agriculture can provide an economic and eco-friendly method of disposal for sustainable agriculture without posing an environmental problem to the groundwater and soil. A field study showed that soil amendment with diluted post-methanation distillery effluent increased the yield of wheat and rice grown in sequence (Pathak et al., 1999). However, the judicious application of PMDE improved crop productivity and alleviated environmental pollution problems (Devarajan et al. 1994; Davamani et al. 2006). Thus, considering the negative impacts of direct discharge of untreated or partly treated effluent on humans and the ecosystem, distillery effluents need to be treated before being discharged into the environment.

4.5 Negative Impact of Biomethanated Distillery Effluent

When the untreated or partially treated melanoidins-containing BMDE are released into water bodies, they damage aquatic ecosystem by reducing the sunlight penetration power and finally the photosynthetic activities and dissolved oxygen content, whereas in agricultural soil it causes depletion of vegetation by

FIGURE 4.3 The potential impact of distillery effluent on the vegetative plants grown in open environment and soil quality (Kumar et al. 2021).

reducing the soil alkalinity and Mn availability and inhibiting the seed germination. The impact of distillery waste on the environment is illustrated in Figure 4.3. In developing countries, BMDE containing recalcitrant compounds is often used to irrigate crops, which adds harmful compounds to agricultural soils. The use of distillery effluent as a source of nutrients can be a viable option due to the presence of some essential minerals, although the presence of several toxic elements can deteriorate soil health. Bharagava and Chandra (2010) also noted that at high concentration, anaerobically treated PMDE acts as an inhibitor for auxin and gibberellin and germination of green gram (*Phaseolus mungo* L.) seeds. Besides, the application of this effluent may cause pollution and changes the nutrient, biological, and biochemical status of the soil, including the composition of microbial communities and enzyme activities. Distillery effluent presents a serious health hazard to the rural and semiurban population that depend on such water bodies for drinking water and other requirements.

5

Treatment Approaches of Distillery Waste for Environmental Safety

5.1 Introduction

Treatment and reuse of wastewater is nowadays a major need all over the world due to the low availability of water resources, as well as the negative impact of the discharge of untreated or partly treated wastewater on the environment (Chandra and Kumar 2015a,b; Kumar et al. 2020a). Distilleries are one of the major ethanol-producing industries; in addition, they are a high consumer of freshwater and utilize the sugarcane molasses as the feedstock for ethanol-making. In terms of ethanol, molasses-based distilleries and fermentation industries have been placed in the "Red" category of most polluting industries by the Central Pollution Control Board (CPCB) and Ministry of Environment, Forests, and Climate Change (MoEF&CC), Government of India. The industrial production of ethanol by fermentation results in the discharge of large quantities of high-strength complex liquid wastes generally termed as spent wash (Kaushik and Thakur, 2009; Mahar et al. 2013; Thiyagu and Sivarajan, 2018). The spent wash yield with respect to the volume of ethanol produced from sugarcane is in the range of 12–15 L of spent wash/liter of ethanol produced. Distillery effluent has also been considered as a threat to the environment in India (Ravikumar, 2015; Kaushik et al. 2018; Kumar and Chandra 2020a; Kumar and Sharma 2019). There are a number of small- and medium-sized industries in India that produce ethanol by fermentation/distillation. In the distillation process, molasses is diluted with water and taken in a fermenter to produce bioethanol with the addition of yeast inoculum. The fermentation broth is filtered and then distilled to concentrate bioethanol. The final effluent after the distillation process is the wastewater that is commonly known as spent wash or vinasses or raw effluent (González et al., 2000; Kumar et al. 2021a; Takle et al., 2018). There are three different organic wastes generated from the molasses-based distilleries, which include yeast sludge, spent malt grain wash, and spent wash. Spent wash is the principal polluted stream generated from distilleries. The physicochemical characteristics of untreated and partially treated distillery spent wash are listed in **Table 5.1**. Spent wash contains 94–97% water and a large amount of inorganic and organic matters, such as potassium, phosphorus, sulfate, calcium, and magnesium (Ferreira et al., 2010; Jain and Srivastava, 2012; Mabuza et al. 2017; Chuppa-Tostain et al., 2018). It is also characterized by a low pH (4.5–4.8), with a typical odor, and dark brown color because of the presence of melanoidins. Due to its nutrient composition, spent wash has been applied in the agricultural sector as fertilizer, mainly in sugarcane crops. However, the discharge of highly colored effluent by these industries causes problems such as environmental damage, health hazards, and aesthetic aspects, as shown in Figure 5.1.

Distillery effluents often contain significant amount of toxic, androgenic and mutagenic organic and inorganic chemical substances persist as environmental pollutants and cross entire food chains providing biomagnification such that organisms at higher trophic levels show higher levels of contamination compared to their prey. In particular, high-molecular-weight, non-biodegradable organic contaminants such as melanoidins, endocrine-disrupting chemicals (EDCs), plant steroids, and polyphenols are among the major sources of toxicants in aquatic environments and highly recalcitrant in the environment (Sangave et al., 2007; Chandra and Kumar 2017a,b; Kumar and Chandra 2020c; Chowdhary et al. 2018a).

DOI: 10.1201/9781003029885-5

TABLE 5.1

The Physicochemical Characteristic of Untreated and Partly Digested Distillery Spent Wash as Reported by Various Researchers

Parameters	Spent Wash			Anaerobically Digested Spent Wash			Distillery Sludge		
	Tiwari et al. (2012)	Ghosh Ray and Ghangrekar (2015)	Thiyagu and Sivarajan (2018)	Mohana et al. (2007)	Pant and Adholeya (2009)	Bharagava and Chandra (2010)	Singh et al. (2014)	Mahaly et al. (2018)	Chandra and Kumar 2017a
Color	Dark brown	–	Dark brown	–	–	Dark black	Dark brown	–	–
Odor	Like molasses	–	Unpleasant	–	–	–	–	–	–
Temperature (°C)	82	–	–	–	–	–	–	–	–
pH	4.2	3.4–3.6	3.8	7.5–8	8.2	8.5	8.52	5.47	8.0
EC (µS/cm)	–	–	45.5	33		–	4.32	0.71	2.292
Total solids (mg/L)	–	34.80	140,260	72,500		47,422	–	–	–
TDS (mg/L)	81,733	7.39	112,400	21,256 (ppm)		17,612	–	–	–
TSS (mg/L)	5,933	23.80	27,860	40,700		29,810	–	–	–
DO (mg/L)	–	–	–	–	–	–	–	–	–
BOD (mg/L)	46,666	28.12	–	8,000–10,000	5,000 (ppm)	12,000	–	–	–
COD (mg/L)	104,130	64.24	162,000	45,000–52,000	25,000 (ppm)	21,000	–	–	–
Total nitrogen [(mg/L)/(mg/kg)]	1,635	–	–	4,284	3.5 (%)	4,096	10.25	0.12 (%)	–
AN [(mg/L)/(mg/kg)]	–	0.61	–	–	–	–		0.12 (%)	190
Total protein [(mg/L)/(mg/kg)]	–	4.25	–	–	–	–	–	–	–
Reducing sugar [(mg/L)/(mg/kg)]	–	3.46	–	0.17 g %	0.23 (%)	–	–	–	–
TOC [(mg/L)/(mg/kg)]	–	–	–	–	–	–	–	62.4 (%)	–
Phosphorus [(mg/L)/(mg/kg)]	163	–	2,100	1,625	–	1,625	8.89	5.63 (%)	–
Potassium [(mg/L)/(mg/kg)]	8,766	–	10,820		2,500 (ppm)	537	5.03	0.98 (%)	–

Parameter								
Sodium [(mg/L)/(mg/kg)]	211	–	–	500 (ppm)	–	–	2.22	56.16
Pehonol [(mg/L)/(mg/kg)]	–	–	7,202	–	6,893	–	–	–
Chloride [(mg/L)/(mg/kg)]	–	10,650	7,997	–	7,842	–	–	1,824.4
Calcium [(mg/L)/(mg/kg)]	1,816	2,975	–	–	–	0.62 (%)	–	–
Sulfate [(mg/L)/(mg/kg)]	1,738	3,015	3,875	–	2,786	–	–	–
C/N ratio	–	–	–	–	–	520 (%)	30.63	–
Lignin (mg/L)	–	–	–	–	–	0.03 (%)	–	–
Trace elements								
Cu [(mg/L)/(mg/kg)]	–	–	–	396 (ppm)	0.75	–	39.2	73.638
Zn [(mg/L)/(mg/kg)]	–	–	–	273 (ppm)	1.24	–	206.0	210.624
Mn [(mg/L)/(mg/kg)]	–	–	–	259 (ppm)	43.63	–	60.5	126.292
Fe [(mg/L)/(mg/kg)]	–	–	–	–	57.50	–	5,292.5	2403.64
Mg [(mg/L)/(mg/kg)]	–	2,380	–	98 (ppm)	–	0.41 (%)	–	21.847
Cr [(mg/L)/(mg/kg)]	–	–	–	–	–	–	–	21.847
Cd [(mg/L)/(mg/kg)]	–	–	–	–	1.30	–	–	1.446
Ni [(mg/L)/(mg/kg)]	–	–	–	–	0.31	–	–	13.425
Pb [(mg/L)/(mg/kg)]	–	–	–	–	0.23	–	–	16.332

BOD: biological oxygen demand; COD: chemical oxygen demand; TDS: total dissolved solid; TSS: total suspended solid; AN: ammonical nitrogen; DO: dissolved oxygen; TOC: total organic carbon; Cu: copper; Zn: zinc; Mn: manganese; Fe: iron; Mg: magnesium; Cr: chromium; Cd: cadmium; Ni: nickel; Pb: lead; EC: electric conductivity.

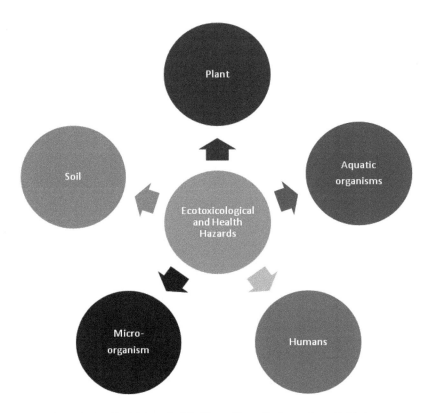

FIGURE 5.1   Ecotoxicological and health hazard of distillery effluent discharges from distilleries.

Various reports have mentioned the direct and indirect toxic effects of the melanoidins-containing distillery effluent, which can lead to the mutagenic effect and chromosomal aberrations besides growth inhibition of plants and different microorganisms (Chandra and Kumar 2017b). It affects the water quality and may become a threat to public health, since certain androgenic-mutagenic compounds or their metabolites are highly toxic and potentially carcinogenic (Chowdhary et al. 2020). It reduces the amount of sunlight to photosynthetic organisms resulting in decreased oxygen levels in aquatic eco-systems. Distillery effluent, if disposed without proper treatment, can cause considerable stress on the water courses leading to widespread damage to aquatic life and reduce soil alkalinity, thereby affecting the quality of groundwater. A reduction in the dissolved oxygen content and eutrophication could also occur in the surrounding water bodies. The direct application of distillery effluent onto soil as fertilizer for sugarcane cultivation could cause soil salinization, the clogging of pores, soil structure destabiliza-tion, and compaction (Kaushik et al. 2005; Sinha et al. 2014; Alves et al. 2015). Due to high pollution rate of effluent, distilleries have been considered as one of the 17 most polluting industries listed by the CPCB. To make them safe for human habitation and consumption, and to protect the functions of the life-supporting ecosystem, spent wash must be properly treated before it is discharged into the environ-ment. Several studies of the removal of melanoidins in connection of distillery effluent decolorization and degradation have been performed using various techniques such as usage of anaerobic and aerobic microorganisms such as fungal, bacterial, algal, enzymatic, and plant (phytoremediation), and various physicochemical methods such as flocculation, ozonation, coagulation, ultrafiltration, electrocoagula-tion, membrane treatment, and advanced oxidation processes (AOPs) such as Fenton, ultraviolet (UV), and UV/H_2O_2 oxidation, etc. (Figure 5.2; Sankaran et al. 2010; Yadav et al. 2011; Solovchenko et al. 2014; Apollo and Aoyi 2016; Santal et al. 2016; Krishnamoorthy et al. 2017; Reis and Hu 2017; Kumar and Chandra 2018a). In physiochemical methods, chemical consumption is extensively very high with huge amount of sludge generation, which requires further safe disposal. In addition, these methods are extremely cost-intensive process as the end products formed during chemical oxidation tends to become

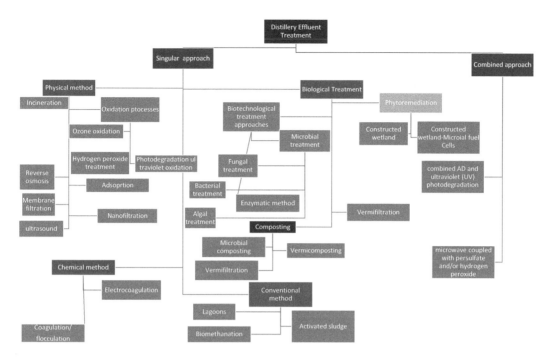

FIGURE 5.2 Schematic representation of various physical, chemical, biological, and integrated approaches used for the treatment of distillery effluent in connection of melanoidin decolorization and degradation.

recalcitrant to total oxidation by chemical means (Prajapati and Chaudhari 2015). A number of biotechnological approaches have been suggested to overcome the problem of physiochemical treatment methods using microorganisms for the treatment of distillery effluent. The effectiveness of microbial decolorization depends on the adaptability and the activity of selected microorganisms (Bharagava and Chandra 2010; Malik et al. 2019). Many microorganisms are capable of degrading melanoidins and melanoidins-containing effluent, including bacteria-activated sludge processes (ASPs), which are largely ineffective in the treatment of distillery wastewaters as a stand-alone process, resulting in little or no color removal from effluent. **Figure 5.2** highlights the common processes involved in treatment of distillery effluent. This chapter focuses on the efficient treatment of distillery effluent by various conventional biological, physical, chemical processes. Further, this chapter also provides an insight into the biotechnological use of bacteria, yeast, fungi, microalgae, and in degradation and decolorization of distillery effluent treatment with an emphasis on environment-friendly phytoremediation technology, which have the ability to remediate the inorganic and organic pollutants present in wastewater and soil. Furthermore, management of distillery effluent using composting techniques is also discussed. Moreover, the impact of various nutritional and environmental parameters on microbial decolorization and degradation of melanoidin-containing distillery effluent is also highlighted in detail. The later part of the chapter discusses the microbial enzyme system involved in the degradation of ECs. Furthermore, the feasibility of distillery effluent for utilization in agriculture filed for ferti-irrigation systems for its management is also discussed.

5.2 Effluent Treatment Plant Scheme

Anaerobic digestion is the best and most practiced first step of effluent treatment plant (ETP) scheme since BOD/COD > 0.25. Pollution load reduction (up to 65% in terms of COD) and energy recovery as biogas are achieved through anaerobic digestion simultaneously (Bhoite and Vaidya 2018b; López et al. 2017). Settling lagoons that are placed after anaerobic digestion are found to be useful to settle the portion of total suspended solids (TSS) by gravity. However, the open lagoons system has several inherent problems of land, odor, and seepage into use and degradation of groundwater. The overflow from settling lagoons is sent to reverse osmosis (RO) plant. The permeate (water) of RO is recycled to the alcohol production unit, reducing the water requirement. The reject of the RO plant is mixed with the press mud and marketed as a biocompost. As alternative to this ETP process, the evaporation technique is followed for treating the distillery wastewater in some distilleries. The outcome of evaporation is pure water and concentrated sludge (approximately 50% moisture). This sludge is further burnt to produce power and the potassium-rich ash is recovered from the combustion of the sludge. Moreover, various technology options are available for the treatment of distillery spent wash. This may include biomethanation, biomethanation and secondary treatment followed by irrigation or disposal in surface water, composting after or without biomethanation, activated sludge treatment, concentration and incineration, anaerobic digestion followed by evaporation and composting, co-incineration, RO, multi-effect evaporators, recovery of potash, or disposal into sea or estuary after or without biomethanation. Anaerobic digestion of distillery spent wash followed by the aerobic oxidation is the common practice in the industries for the distillery spent wash treatment (Jiménez et al. 2003; Reis et al. 2019). However, the effluent after conventional anaerobic digestion still retains high COD (40,000 mg/L), dark brown color, and becomes recalcitrant with low biodegradability index (BI: $BOD_5/COD < 0.2$). The post-anaerobic effluent has high BOD in the range of 5000–7500 mg/L, which is required to be brought below 100 mg/L for disposal on land or below 30 mg/L for disposal in water bodies through secondary treatment. It is impossible to get BOD up to 30 mg/L through biological treatment whatsoever may be the treatment system. Probably it is due to the fact that the most difficult biodegradable portion is left out at the end of the treatment (Blonskaja and Zub 2009; López et al. 2017). Further, the ASP is generally used to reduce the BOD and COD of post-anaerobic effluent (Ahansazan et al. 2014). Subsequent technologies, from biomethanation to co-incineration, which utilize the energy-generating, irrigation, and fertility potential of spent wash, have improved the management of distillery spent wash to change it into an environment-friendly, socially acceptable, zero effluent discharge industry. The direct discharge of biomethanated distillery wastewater onto the land causes seed germination inhibition and reduces soil alkalinity; while discharge into river diminishes the penetration of light and reduces dissolved oxygen level in the water (Kumar and Gopal 2001; Kaushik et al. 2005; Ramakritinan et al. 2005; Ayyasamy et al. 2008; Chauhan and Rai 2010). The dark-brown-colored wastewater after anaerobic treatment and dilution is used for irrigation causing gradual soil darkening, which affects the physicochemical properties of soil and is also harmful for soil microorganisms (Chauhan and Rai 2010; Sinha et al. 2014). Sometimes, the leaching of spent wash into the groundwater table resulted in severe groundwater contaminations. Therefore, it is necessary to study additional pre- or posttreatment methods to remove color and recalcitrant organic compounds from distillery effluent. Tertiary treatment, comprising physicochemical methods, adsorption, and advanced chemical oxidation processes, basically adopted for removal of color in addition to tracing organics, involves high operational cost (Rodriguez-caballero et al. 2012; Apollo and Aoyi 2016; Kumar et al. 2020a). However, these methods and strategies are less efficient in removing pollutants of wastewater due to its recalcitrant nature and also chemical consumption is very high with a huge amount of sludge generation, which requires further wastewater treatment before its safe disposal (Kumar et al. 2021a). **Figure 5.3** shows the process flow of common effluent treatment plant for distillery effluent treatment and management.

5.2.1 Conventional Treatment Technologies for Distillery Effluent

Distillery effluents may be treated by a number of methods, either singly or in combination. The conventional wastewater treatment processes are designed to reduce solids in suspension, biodegradable organic products, microorganisms, and nutrients, but not to reduce the chemical pollutants and much less emerging chemical pollutants (Sangave et al. 2007; Rani et al. 2013). However, for the treatment of spent wash by conventional methods like aerobic and non-aerobic digestion, the ratio of BOD to COD should be 40.6. Conventional anaerobic-aerobic wastewater treatment processes can remove only

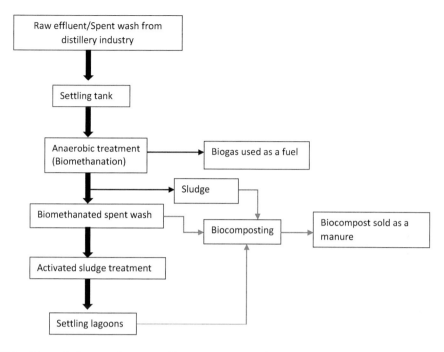

FIGURE 5.3 Schematic diagram of common effluent treatment plant for waste management.

6–7% maximum of the initial concentration of melanoidins. The following anaerobic technologies used for effluent treatment will be discussed in detail next:

  i. Methane recovery processes (biomethanation)
 ii. Anaerobic lagooning
iii. Activated sludge treatment

5.2.1.1 Methane Recovery Processes (Biomethanation)/Anaerobic Treatment

Biomethanation is the most popular and widely accepted form of treatment of distillery effluent adopted in distilleries. Almost 80–90% of distillery ETPs in India has anaerobic reactors used for effluent treatment. Biogas generated through the anaerobic reactors made the system pay back for its cost within 2–3 years. Although these methods are employed to remove the color, their capacity to reduce the toxicity is still a matter of major concern (Singh et al. 2010; Bhoite and Vaidya 2018a). The effluent received after anaerobic digestion is called post-methanated distillery effluent (PMDE), anaerobically digested spent wash, or biomethanated distillery effluent (BMDE). PMDE is characterized by an extremely higher level of BOD, COD, TDS, sulfate, phenol, potassium and phosphorous with alkaline pH, and dark brown color (Pant and Adholeya 2009; Sarwat Chandra et al. 2014; Wagh and Nemade 2018; Malik et al. 2019). In addition, PMDE retains a high amount of various heavy metals along with melanoidins and various recalcitrant organic compounds such as pyrroline, 3-hydroxypropanal, butanoic acid, 2- methylhexane, 5-methyl-2-furancarboxaldehyde, 2-methoxyphenol, 5-(hydroxymethyl)-2- furfural, *n*-methyl indane, 4-methyl guaiacol, *p*-chloroanisole, 4-vinyl-2- methoxyphenol, 2,6-dimethoxyphenol, 2-nitroacetophenone, dihydroxy coniferyl alcohol, *trans*-2-tridecenal, 3-heptyl-5-methyl-2,3h-furanone, and benzyl-3,4-ethylenedioxypyrrol-2,5-dicarboxylate, etc. Although anaerobic treatment of distillery spent wash is feasible and quite appealing from an energy point of view, the presence of refractory organic compounds can be toxic or inhibitory for anaerobic microorganisms. This slows down the kinetics and reduces mean rates of methane production and yield of effluent treatment (Acharya et al. 2011; Bhoite and Vaidya 2018b). Even though these procedures prove to be efficient, the operational costs are relatively high and lead to other disadvantages like sludge formation and biomass accumulation and the inability to treat a wide array of Maillard products having structural diversity may even have secondary pollution problems (Sa et al. 2008; Sinha et al. 2014). Moreover, the high organic content, variable composition, and the large seasonal flow variations of spent wash may result in operational difficulties associated with treatment. Biomethanated distillery effluent can be used as an organic fertilizer as it is characterized by high potassium and trace metal content that facilitate the growth of crops (Kaushik et al. 2005; Chauhan and Rai 2010). But the effluent cannot be disposed onto the land directly because it inhibits seed germination and reduces soil alkalinity as well as manganese availability, whereas the discharge of spent wash into water bodies in rivers, lakes, and lagoons and streams releases an unpleasant odor and causes reduction in penetration of sunlight, which in turn decreases both dissolved oxygen concentration and photosynthetic activity of marine plants, creating anaerobic conditions that kill most of the aquatic life (Juwarkar and Dutta 1989; Kumar and Gopal 2001; Ayyasamy et al. 2008; Alves et al. 2015). Direct dumping of such waste posed a risk of soil salinization and contamination of groundwater by heavy metals. Hence, the discharge of biomethanated effluent is ecotoxic and unsafe to use in any form for humans and environment such as agricultural and aquatic ecosystems (Jain et al. 2005; Pandey et al. 2008; Kumar and Sharma 2019; Kumar and Chandra 2020a). A BOD:N:P ratio of 100:2.4:0.3 suggests that anaerobic treatment methods at the primary stage were followed by aerobic treatment methods for reducing the pollution potential of distillery effluent. For the efficient treatment of raw spent wash, there are various technologies that utilize resource recovery and disposal:

1. Biomethanation followed by solar evaporation
2. Biomethanation followed by two-stage aerobic treatment irrigation
3. Biomethanation followed by two-stage aerobic treatment
4. Tertiary treatment followed by surface water discharge
5. Biomethanation followed by evaporation and incineration potassium recovery

5.2.1.2 Anaerobic Lagooning

Anaerobic lagoons are one of the effective and a preferable choice for anaerobic treatment of distillery waste. Rao (1972) had done excellent research work in the field of distillery waste management by using anaerobic lagoon treatment in two pilot-scale lagoons in series and obtained overall BOD removal ranging from 82% to 92%. However, the lagoon systems are seldom operational, souring being a frequent phenomenon. The treatment of spent wash in lagoons generates greenhouse gas emissions, and ferti-irrigation practices may in some cases negatively affect the structure of soils and aquifers.

5.2.1.3 Aerobic Treatment: Activated Sludge Process (ASP) System

Biological treatment using aerobic ASP has been in practice for well over a century. The ASP was developed in the early 1900s for the treatment of domestic wastewaters, and it has since been adapted for removing biodegradable organics from industrial wastewaters. This is the most common and oldest biotreatment process used to treat municipal and industrial wastewater (Ahansazan et al. 2014; Acharya et al. 2011). Biological treatment by activated sludge offers high efficiency in COD reduction, but does not completely eliminate the color of the wastewater due to the biodegradable difficulty of the melanoidins presented in the effluents (Sales et al. 1987). The ASP mineralizes organic compounds into carbon dioxide and water and generates a huge amount of bacterial biomass that is the predominant component of so-called secondary sludge. On average, the ASP produces about 0.4 g biomass per 1 g of COD removed. The wastewater after primary treatment, i.e., suspended impurities removal, is treated in an ASP-based biological treatment system comprising aeration tank followed by secondary clarifier. The ASP uses a mass of microorganisms (usually bacteria) for aerobic treatment of effluent (Satyawali and Balakrishnan, 2008). Organic contaminants in the effluent provide the carbon and energy required to encourage microbial growth and reproduction; nitrogen (urea) and phosphorous (phosphoric acid) are sometimes added when the raw wastewater is an insufficient source of required nutrients to promote this growth. The effluent is thoroughly mixed with air in an aeration tank, and the organic matter is converted into microbial cell tissue and carbon dioxide. The microbial mass together with the effluent is referred to as a mixed liquor. After a specified time in the aeration tank, the mixed liquor passes into a settling tank, or clarifier, where the biomass, or sludge, settles by gravity from the treated effluent. This continuous process produces an effluent that may be treated further or discharged directly to a publicly owned treatment works. In addition, ASP generates large amounts of low-value bacterial biomass. The treatment and disposal of this excess bacterial biomass, also known as waste-activated sludge, accounts for about 40–60% of the wastewater treatment plant operation cost. Before it enters the aeration basin, raw wastewater (influent) is normally pretreated to remove easily settled solids or other suspended matter. The pretreated influent is pumped into the aeration tank (or basin), where it is combined with the microorganisms. In order to maintain the bacterial mass in the aeration basin, aerators perform two important functions: they keep the liquid and sludge agitated, and they promote the transfer of oxygen into the wastewater. The aerators run continuously and are strategically positioned to keep the mixed liquor in suspension throughout and avoid the formation of dead zones. Foam generated by the aerators can be contained by anti-foam sprays. The pH of the incoming wastewater or mixed liquor must be steadily controlled for proper microorganism growth. A neutral pH is maintained by adding a caustic or acid, usually 25% by weight in solution. In colder climates, supplemental heat is required to maintain an acceptable temperature range for the microorganisms. Indirect heat is added to the base of the aeration basin via a steam bayonet or a gas-fired heater. Some plants save energy by first preheating the influent with a hot process stream (e.g., distillation column bottoms) in a heat exchanger. After spending a specified amount of time in the aeration basin, water overflows into a clarifier, where the microbial solids settle by gravity and are separated from the treated water. The clarifier consists of a tank with a center feed well, which promotes even dispersion. The settled sludge is raked from the quiescent zone at the bottom of the clarifier with mechanical scrapers. A portion of the biomass is recycled back to the aeration basin to maintain the desired concentration of organisms in the basin. The remaining sludge is pumped from the system, typically by a motor-driven centrifugal pump equipped with a flow meter, for final disposal. Floating materials that collect at the surface of the liquid in the clarifier are collectively referred to as scum. A skimmer constantly moves the scum toward a collection trough that drops it into a receiver for later removal. Both the sludge scraper and the scum skimmer are connected to a common drive shaft powered by a small motor. Conventional treatment, generally a combination of aerobic and anaerobic processes, normally degrades melanoidins by only 6% or 7%. Increasing pressure to meet more stringent discharge standards or not being allowed to discharge treated effluent has led to implementation of a variety of advanced biological treatment processes in recent years. A conventional activated sludge system for treatment of raw distillery effluent is presented in **Figure 5.4**.

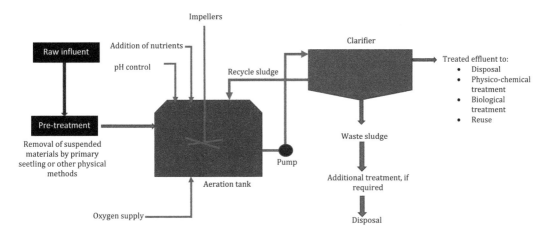

FIGURE 5.4 A conventional activated sludge system for treatment of raw distillery effluent.

5.3 Physicochemical Treatment

5.3.1 Concentration and Incineration

It was recognized in 2001 by the CPCB that concentrating or drying the distillery effluent and burning (incineration) with ancillary fuel, with energy recovery in the form of steam, is the most attractive alternative as it leaves a small amount of residue, which could be used as a fertilizer. During concentration and incineration, the large quantum of spent wash is reduced by increasing the viscosity and then finally it is dried as ash in furnace chambers to let out as solid material. Concentrated spent wash at 55–60% solids or spent wash powder can be used to run a specially designed boiler with or without subsidiary fuel. Generated steam can be used to run a steam turbine to generate electricity and exhaust steam can be used for distillery and evaporation plant operation. Concentration and incineration systems are economically better for distilleries beyond 60 KLPD capacity. The Charter on Corporate Environmental Responsibility for Environmental Protection recognizes this as one of the suggested measures prescribed to be used in combination. Importantly, the CPCB does not rely on any one measure totally and prescribes that molasses-based distilleries will ensure compliance with any combination of methods. Only a small number of distilleries, about 27, have adopted concentration and incineration process for spent wash treatment. In Uttar Pradesh, there are just three distilleries that have adopted incineration. The co-incineration initiative of the CPCB involves incineration of concentrated spent wash as fuel in cement/steel industries along with other fuel/raw materials. This facilitates destruction of wastes at higher temperatures of 1200–1400°C and incorporation of the inorganic contents of the pollutants in the clinker. Concentration of spent wash and its usage as an animal feed additive is a common practice among countries producing alcohol from beet molasses (Europe, North America). This practice has not found acceptance because Indian spent wash contains a higher percentage of inorganic substance that produce a laxative effect if the consumption of feed is not closely monitored. Besides, in Indian conditions, the cost of concentration of spent wash is prohibitive. This process of concentration and incineration is at present still in a trial phase and is very expensive to adopt with high capex and also high running cost that raises the cost of production of alcohol exorbitantly. In addition, this process creates problems of scaling, choking of boilers, frequent failing of furnace refractory material that affects bed fluidization, heavy sludge accumulation, and frequent stoppages of the plant for descaling, which hampers not only the production process but also the effluent treatment process as well. Apart from this installation of special type of boiler that needs to be replaced every 3–5 years due to high wear and tear and quick erosion, disposal of ash after effluent incineration and loss of biogas are major concerns associated with this technology and the sustainability of this technology may still need to be established. Concentration and incineration technologies however present problems of disposal of sludges from storage tanks, condensates from evaporators, and rejects from the RO plant.

5.3.2 Solar Evaporation

Solar evaporation has been recognized as an effective applications of solar for wastewater treatment. However, this technique requires a large land area and also needs to take into consideration the weather conditions prevailing in the region. It is also non-functional during the monsoon.

5.3.3 Adsorption

Adsorption, also known as bio-oxidation, is a wastewater purification technique for removing a wide range of organic and inorganic compounds from wastewater. It is most commonly implemented for the removal or low concentrations of non-degradable organic compounds from groundwater, drinking water preparation, process water, or as tertiary cleansing after, for example, biological water purification (Barakat 2011; Prajapati and Chaudhari 2015; Fito et al. 2019). Moreover, the adsorption is used for treatment of distillery effluent with high BOD, COD, TDS, TSS, and heavy metals. In recent years, the use of dry biomass (biosorption) for the removal of organic and inorganic pollutants has gained credibility because of its eco-friendly nature, excellent performance, and low cost (Kaushik and Thakur

2013). Ohmomo et al. (1988) conducted an experiment to elucidate the mechanism of adsorption of melanoidin to the mycelia of *Aspergillus oryzae* Y-2-32. *A. oryzae* Y-2-32 decolorized about 75% of a melanoidin solution when it was cultivated at 35 °C for 4 days on glycerol-peptone medium with shaking. This fungal strain shows melanoidin decolorizing activity due to the adsorption of melanoidin to mycelia. Further in 1995, Sirianuntapiboon and his colleagues elucidated the absorption mechanism for decolorization of melanoidins by fungal strain *Rhizoctonia* sp. D-90. *Rhizoctonia* sp. D-90 decolorized molasses-melanoidin medium and a synthetic melanoidin medium within 8 days by 87.5% and 84.5%, respectively, under experimental growth conditions. Authors found that mycelia grown in solutions of melanoidin turned dark brown and remained so after being washed several times with buffer solution. Further, the transmission electron microscopy showed that the mycelia absorbed melanoidin pigment and accumulated copious electron-dense materials in the cytoplasm exclusively, when *Rhizoctonia* sp. D-90 was grown in melanoidin medium. The electron-dense materials in the cytoplasm were very probably composed of melanoidin pigment that had been absorbed by the mycelia. According to Sirianuntapiboon et al. (1995), the mechanism for decolorizing melanoidin first involves the absorption of melanoidin pigments by the mycelial cells in the middle of growth, and then their accumulation intracellularly in the cytoplasm and near the cell membrane. The melanoidin pigments are decolorized by the active oxygen produced by the reactions with sugar oxidase intracellular decolorizing enzymes, and the electron-dense materials disappear from the mycelia without any change in the color of the culture filtrate during culture in fresh colorless medium or during prolonged cultivation in melanoidin-containing medium. Probably the melanoidins in adsorbed into cells as a macromolecule acts as a peroxidase-scavenger. Similarly Ziaei-Rad et al. (2020) studied the bioadsorption and enzymatic biodecolorization of effluents discharged from ethanol production plants using *Aspergillus fumigatus* UB$_2$60. *A. fumigatus* UB$_2$60 removed 47% of color at an incubation time of 96 h at 30 °C. Fourier transform infrared spectroscopy (FTIR) studies of *A. fumigatus* UB$_2$60 confirmed that the biomass surface changed considerably before and after bio-decolorization. Furthermore, the ultramorphological results indicated that *A. fumigatus* UB$_2$60 underwent considerable structural changes in the presence of sodium hydroxide, which absorb the color from the fungal surface. The authors recommended a detailed analysis of the fungus-melanoidin interaction is required to understand the mechanism and percentage of each process (biosorption, biodegradation, and bioaccumulation) in color removal from distillery wastewater. Comparative studies of color removal from distillery effluent using bagasse fly ash and commercially activated carbons (CAC) showed 58% color removal with 30 g/dm^3 of bagasse fly ash and 80.70% color removal with 20 g/dm^3 of CAC (Satyawali and Balakrishnan 2007). *Leucaena leucocephala* was reported as a potential material for low-cost adsorbent due to their properties. The potential of activated carbon produced from *L. leucocephala* charcoal and application in melanoidin removal has been studied by Insoongnoen et al. (2020). Maximum adsorption capacity of melanoidin on the *L. leucocephala*-activated carbon (LAC) samples were found in the range of 588.24–666.67 mg/g. The optimum adsorption conditions of the LAC were obtained at the contact time of 480 min, the initial melanoidin concentration of 1000 mg/L, initial solution pH of 2.0, and agitation speed of 100 rpm at the temperature of 65 °C. Similarly, Ahmed et al. (2020) utilized fly ash as a low-cost adsorbent and a major coal-combustion by-product generated in thermal power plants in huge quantities to reduce melanoidin from distillery effluent. The maximum melanoidin removal efficiency of 84% was achieved under optimized conditions, namely, initial concentration of 1100 mg/L (5% dilution), pH 6.0, and a contact time of 120 min. Further, the maximum adsorption capacity of 281.34 mg/g was observed using the Langmuir isotherm. Rafigh and Soleymani (2019) reported the melanoidin removal from molasses wastewater using graphene oxide nanosheets (GONs) as a cost-effective, multifunctional material, derived from graphene by oxidative treatment, and an efficient and potential adsorbent for melanoidin removal. Microwave coupled with persulfate and/or H$_2$O$_2$ system is an effective technique for melanoidin removal (Tripathy et al. 2020). This adsorption process was reported to have achieved more than 90% of melanoidin removal. However, because its high operating cost has limited its use for wastewater treatment, alternative low-cost adsorbents are being found from various materials such as plant waste. In recent years, the use of plant biomass for the removal of organic and inorganic pollutants from wastewater has gained credibility due to its environment-friendly nature, low cost, and excellent performance. Biomass from invasive macrophytes, such as *Pistia stratiotes* and *Eichhornia crassipes*, has been reported as being highly adsorbent for removal of color from anaerobically digested sugarcane stillage (Sánchez-Galván et al., 2015). The FTIR spectra analysis suggested the electrostatic attraction between protonated carboxylic groups on biomass and anionic colored compounds as being one of the adsorption mechanisms for stillage decolorization.

5.3.4 Biosorption

Biosorption for metal removal is found to be an attractive technology as it is inexpensive, with high efficiency and specificity, less chemical or sludge formation, no additional nutrient requirements, regeneration of biosorbents, and eco-friendly in nature. Recently, this technology has been receiving huge attention in research and industries as an emerging technology for wastewater pollutants removal (Barakat 2011; Kidgell et al. 2014; Kumar et al. 2020a). Biosorption is a physiochemical process that occurs naturally in certain biological materials (biomass), which allows it to passively concentrate and bind contaminants onto its cellular structure. Biosorption can be defined as the ability of biological materials to accumulate heavy metals from wastewater through metabolically mediated or physico-chemical pathways of uptake. This method has been studied by several researchers as an alternative technique to conventional methods for heavy metal removal from wastewater (Mani and Kumar, 2014). Different types of biosorbents such as activated carbon (Liakos and Lazaridis, 2016;), chitosan, chitin, alumina, silica gel, zeolite, sawdust, rice husk, maize cobs, fly ash, red mud, and bagasse pith are being used to remove melanoidins from distillery effluent. The removal of melanoidins from simulated and real (untreated and biologically treated) molasses effluents was investigated by adsorption onto powdered activated charcoal (PAC) in batch adsorption experiments (Liakos and Lazaridis, 2016). The maximum experimental adsorption capacity was found to be approximately 10–12 g/g. Melanoidin-loaded PAC can be regenerated by sodium oleate, providing reasonable melanoidin recovery for seven successive cycles of adsorption-desorption. The performance efficiencies of powdered activated carbon and granular activated carbon (GAC) were compared toward the degradation of organic pollutants in terms of color (David et al. 2016). Optimization of the process parameters using Taguchi Orthogonal Array (OA) experimental design resulted in a maximum of 95% spent wash decolorization using PAC and 44% using GAC. Using Artificial Neural Network (ANN), a two-layered feedforward backpropagation model, resulted as the best performance and predictive model for spent wash decolorization. Biosorbents are simple, low-cost, flexible, and energy-saving and have high pollutant removal efficiency. Microorganisms such as algae, fungi, yeast, molds, and bacteria have been used for the recovery of heavy metals from the metal-contaminated environment. Vivekanandam et al. (2019) prepared an ingenious bioorganic adsorbents for the removal of distillery-based pigment-melanoidin and further studied their adsorption mechanism. In this study, deportations of melanoidins were investigated using both natural adsorbents and constructed microbial consortia-based adsorbents and denoting it using the adsorption isotherms and scanning electron microscopy analysis. Based on the performances, both microbial-coated commercially activated carbon (MCCAC) and CAC were better when compared with the other adsorbents at lower melanoidin concentrations; however, at higher melanoidin concentration, MCCAC performed better. Rizvi et al. (2020) introduced a holistic approach for melanoidin removal via Fe-impregnated activated carbon prepared from *Mangifera indica* leaves biomass. With the adsorbent dose (62.5 mg/L), initial melanoidin concentration (550 mg/L), temperature (40°C), contact time (75 min), and Fe-adsorbent ratio (30%), the maximum melanoidin removal (85.6%) was achieved from waste leaf biomass of *M. indica*, depicting its potential for the melanoidin removal from wastewater. Although the use of dry biomass for the removal of organic and inorganic pollutants has excellent results, the disposal of solid adsorbents after treatment itself is a big problem because they contain toxic compounds on their surfaces. In addition, their application is limited as biological treatment requires a large land area, has sensitivity toward toxicity of certain chemicals, and treatment time is very high.

5.3.5 Forward Osmosis

Forward osmosis (FO) is a membrane-based separation process operating on osmotic pressure difference between the low osmotic pressure feed solution and the high osmotic pressure draw solution separated by a semipermeable membrane. In engineered FO process, a semipermeable membrane separates the feed and draw solutions. Water from the lower osmotic pressure feed passes through the membrane to the higher osmotic pressure draw solution, leading to the concentration of the feed and dilution of the draw solutions (Kehrein and Garfí 2020). In combination with other membrane separation processes like RO, membrane distillation, and microfiltration, FO has been used for treatment of various complex wastewaters to either enrich the feed in trace components by reducing the feed volume or to reclaim the wastewater for direct potable reuse. Singh et al. (2019) treated high-strength wastewater of sugarcane molasses distillery by FO using aquaporin biomimetic membranes and magnesium chloride hexahydrate ($MgCl_2 \cdot 6H_2O$) as draw solution. The operational parameters, namely, feed solution and draw solution flow rate and draw solution concentration were optimized using 10% v/v melanoidins model feed solution. Under the conditions of this work, feed and draw flow rates of 1 L/min and draw solution concentration of 2 M $MgCl_2 \cdot 6H_2O$ for melanoidins model solution and 3M $MgCl_2 \cdot 6H_2O$ for distillery wastewater were optimal for maximum rejection. Rejection of 90% melanoidins, 96% antioxidant activity, and 84% COD was obtained with melanoidins model feed, with a corresponding water flux of 6.3 L/m² h. With as-received distillery wastewater, the rejection was similar (85–90%) to the melanoidins solution, but the water flux was lower (2.8 L/m² h). Water recovery from distillery wastewater over 24 h study period was higher with FO (70%) than reported for RO (35–45%). Repeated use of the FO membrane over five consecutives 24 h cycles with fresh feed and draw solutions and periodic cleaning showed consistent average water flux and rejection of the feed constituents. Subsequently, with stringent environmental regulations applied on reclamation of wastewater for reuse, FO is faced with major challenges, such as high energy consumption and high requirement of a second treatment process to treat the secondary pollutants (sludge).

5.3.6 Coagulation-Flocculation

Coagulation-flocculation is an essential unit process in removing colloidal particles and natural organic matters (NOM) from water and wastewater. Since melanoidins are chemically similar to NOM, this process is possibly applicable in removing melanoidins (Liang et al. 2009). Physicochemical treatment allows reducing dissolved, suspended, colloidal, and non-settleable materials from water or wastewater through chemical coagulation-flocculation followed by gravity settling. In water and wastewater settings, coagulation is being considered as the first step of a physicochemical treatment process. This involves the addition of coagulants to subvert colloidal impurities and coalesce dissolved organic materials (DOM) and small particles to large aggregates. These become unstable and due to gravity tend to settle. These aggregates are commonly removed by sedimentation, filtration, or flotation mechanisms. Coagulation consists of neutralizing the negative surface charges of colloidal particles, while flocculation is the aggregation of neutralized-particles followed by floc formation. In coagulation, the electrostatic attraction between oppositely charged soluble dye and polymer molecule coagulates the effluent. So far, two categories of coagulants—synthetic (mainly inorganic) and natural/plant based—have been discussed in literature for coagulation-flocculation-based distillery effluent treatment (David et al. 2016). The first one covers mainly ferric chloride ($FeCl_3$) (Liakos and Lazaridis 2014; Zhang et al. 2017), aluminum chlorohydrate (ACH), and polydiallyldimethylammonium chloride (polyDADMAC) (Fan et al. 2011), Alum ($Al_2(SO_4)_3 \cdot 16H_2O$), and calcium oxide (CaO) (Nguyen et al. 2010), which are extensively used in distillery wastewater treatment. Coagulation removal of melanoidins and COD from biologically treated molasses effluent using ferric chloride has been widely studied (Liang et al. 2009a). Under the optimum conditions, up to 86% and 96% of COD and color removal efficiencies were achieved while the residual turbidity in supernatant was less than 5 NTU and Fe^{3+} concentration was negligible because of effective destabilization and subsequent sedimentation. Liang et al. (2009b) reported the efficient treatment of molasses wastewater by coagulation/flocculation. Authors indicated that ferric chloride was the most effective among the conventional coagulants, achieving 89% COD and 98% color eliminations; while aluminum sulfate was the least effective, giving COD and color reductions of 66% and 86%, respectively. In addition to metal cations, counterions exert significant influence on the coagulation performance since Cl^--based metal salts attained better removal efficiency than SO_4^{2-}-based ones at the optimal coagulant dosages. The performance of a coagulation sequence using ACH and a low MW polyDADMAC and $FeCl_3$, for decolorizing a high-strength industrial molasses wastewater, was compared at bench scale (Fan et al., 2011). At their optimum dosages, ACH/polyDAD-MAC gave higher color removal than FeCl$_3$ (45% cf. 28%), whereas COD reduction was similar (~30%), indicating preferential removal of melanoidins (a major contributor to the color) by ACH/polyDADMAC. Anaerobic biotreatment of the wastewater enhanced the coagulation efficiency markedly, with FeCl$_3$ achieving 94% color and 96% COD removal, while ACH/polyDADMAC gave 70% and 56% removal, respectively Liakos and Lazaridis (2014) examined the removal efficiency of melanoidins from simulated and real wastewaters by coagulation and electroflotation processes. The results show that coagulation experiments achieved color removal 90% and higher, at pH 5.0, for all wastewaters but with different ferric ion dose. Real effluents could be discolored by 100 mM [Fe^{3+}], while simulated by 300 mM [Fe^{3+}]. After flocculation, the generated ferric hydroxide sludge was washed, solubilized, and reused effectively in a new run. Melanoidin removal was also studied by electroflotation. Color removal was 95%, 90%, and 45% for real treated, real untreated, and simulated wastewaters, respectively, by applying 0.5 A current intensity. Furthermore, coagulation can reduce significantly the COD content of real effluents. The decolorization performance was dependent on pH value since the lower pH values led to faster reactions and higher decolorization efficiency. Some of the notable drawbacks of such coagulants include high operational cost, toxic and/or non-biodegradable voluminous sludge generation, and reduced efficiencies at low temperature. On the contrary, the natural coagulants (polymeric macromolecules derived from plants/microorganisms/animals and having polysaccharides or proteins as coagulating agent) have been proven of great ecological and environmental importance for removal of melanoidins from wastewater (Prasad and Srivastava 2009; David et al. 2015). The inevitable features of such coagulants include their biodegradability, wide sources, rich variety, unlikely to alter the pH, and conductivity of treated water. It has been reported that coagulation-flocculation is affected by several parameters such as coagulant dose, pH of the growth medium, properties of the cellular surface, cell density, coagulant type and dosage, time, the type of effluent, concentration of pigments, and other processing aids used (Prasad et al. 2009; Zhang et al. 2017). The pH of a culture is one of the most important factors that affect flocculation because it determines the net surface charge (positive, negative, or even neutral) present on the microalgae cells. Furthermore, the ionization of certain functional groups at the cell surface, such as carboxyl and amino groups, is highly pH dependent and has a significant impact on the physical-chemical characteristics of microbial cells. Momeni et al. (2018) used chitosan/(3-chloro-2-hydroxypropyl)trimethylammonium chloride (CHPATC) as coagulant to remove color and turbidity of industrial wastewater. To achieve the optimum conditions for the removal of color and turbidity, the response surface methodology experimental design method was used. This study showed that the optimum conditions for the removal of color and turbidity were 76.20% and 90.14%, respectively, indicating the high accuracy of the prediction model. This method can completely eliminate the color, but has the main drawbacks of the difficult management of huge amount of sludge produced after coagulation-flocculation, the high chemical costs, as well as the low removal efficiency of soluble COD. However, coagulation results in generation of large amounts of sludge and total dissolved solids in the effluent are further increased.

5.3.7 Electrocoagulation

Electrocoagulation (EC) is a proven simple and cost-effective technology that has been demonstrated to be suitable for removal of heavy metals, including copper, from wastewater streams in a wide range of industrial sectors. Kobya and Gengec (2012) investigated the decolorization of melanoidins by electrocoagulation process using aluminum electrodes/EC processes that were extremely efficient and able to achieve a decolorization efficiency of >98% at pH $i = 4.2, j = 5$ A/m^2, $\kappa = 2500$ µS/cm, CO = 100 mg/L, and $t_{EC} = 10$ min. Jack et al. (2014) explored the role of EC for the removal of copper from distillery waste streams. Copper could not be removed from caustic wash water, as passivation of the electrodes meant that no floc was formed. However, the wash waters could be treated if mixed with spent lees, with an 80% reduction in copper being obtained. The electrocoagulation system was scaled up and its performance evaluated in a trial at a large Scotch malt whisky distillery. Copper reductions of 88% were achieved at low-power consumption (34 Wh/m^3), while at 112 Wh/m^3 residual copper levels were reduced by 96%. This trial was carried out at a flow rate of 1000 L/h, demonstrating that the technology could readily handle the volumes and flow rates required in practice. Both the capital and running costs of an electrocoagulation system are low, while the technique presents other advantages over the existing copper removal technologies currently in use in the distilled spirits sector. Asaithambi et al. (2021) analyzed the influence of direct and alternating current on electrocoagulation process in terms of % color and % COD removal along with electrical energy consumption from distillery industrial effluent. The percentage of color and COD removal and electrical energy consumption were about 90.57%, 86.54%, and 3.50 kWh/m^3, respectively, with direct current electrocoagulation (DCE). For alternating current electrocoagulation (ACE), it was 100%, 95%, and 3.20 kWh/m^3, respectively, at the optimal experimental condition of COD—3000 mg/L, initial wastewater pH—7, current density—0.4 A/dm^2, interelectrode spacing—1 cm, combination of electrode—Fe/Fe, pulse duty cycle—0.45, frequency—50 Hz, and treatment time—3.5 h.

5.3.8 Electrochemical Treatment

Electrochemical treatment processes can provide valuable contributions to the protection of the environment through the minimization of waste and toxic materials in effluents. This treatment is widely used to remove the color from industrial effluent (Prajapati et al. 2016). Electrochemical processes use electron as main reagent to the effluent to convert chloride to chlorine/hypochlorite, but also require the presence of supporting electrolytes. In this process, the pollutants are destroyed by either the direct or indirect oxidation process. In a direct anodic oxidation process, the pollutants are first adsorbed on the anode surface and then destroyed by the anodic electron transfer reaction. In an indirect oxidation process, strong oxidants such as hypochlorite/chlorine, ozone, and hydrogen peroxide are electrochemically generated and pollutants are then destroyed in the bulk solution by oxidation reaction of the generated oxidant. Among the oxidants, hypochlorite is cheaper and most of the effluents have a significant amount of chloride. Manisankar et al. (2003) reported the complete color removal of industry-treated distillery effluent through an electrochemical technique. In this treatment system, two different types of anodes, planar graphite (Gr) and titanium substrate insoluble anodes (TSIA), were chosen for the treatment of distillery effluent. Lead dioxide coated on titanium (PbO$_2$-Ti) and ruthenium oxide coated on titanium (RuO$_2$-Ti) electrodes were used as TSIA. Current density (CD) was varied from 1.5 to 5.5 A dm^2. Complete decolorization was obtained in both cases. A maximum of 92% of COD reduction, 98.1% of BOD reduction, and 99.5% of absorbance reduction were obtained in the setup in which RuO$_2$-Ti as anode and stainless steel as cathode were used. A probable mechanism has been proposed for the oxidation of organics present in the effluent. Prasad and Srivastava (2009) conducted an electrochemical degradation experiments with ruthenium-oxide-coated titanium mesh acting as anode and stainless steel as cathode to degrade distillery spent wash. The optimal removal of color of 83.31% and COD degradation of 39.66% was obtained at current density of 14.285 mA/cm^2), electrolysis time of 3 h, and at dilution of 10% distillery spent wash at slightly acidic pH 5.5. The actual color removal and COD degradation at optimal conditions are 81% and 37%, respectively, which confirms close to factorial design results. Asaithambi et al. (2012) studied the influence of experimental parameters in the treatment of distillery effluent by electrochemical oxidation. The maximum removal of COD was observed to be 84% at a current density of 3 A/dm^2 at an electrolyte concentration of 10 g/L with an effluent COD concentration of 1000 ppm and at an initial pH of 6.0. The ozonation of a chemical industry wastewater contains many complex organic pollutants and presents high chloride content and toxicity (Lima et al. 2006). The electrochemical methods for the treatment of dye effluent can be significant since the final products are not hazardous; however, it requires more input of power cost. These methods are expensive; inefficient in the color removal, and significant amount of sludge generation results in pollution transfers, which are the major bottlenecks for its implementation in treatment processes.

5.3.9 Membrane Filtration

Membrane filtration technology, as an advanced separation technology, has gained popularity and plays a major role in the advanced wastewater treatment systems because of its potential for recovery of wastewater from many industries. In membrane filtration, the appropriate membrane is capable of removing all types of organic and inorganic pollutants. Membrane processes such as microfiltration (MF), ultrafiltration (UF), nanofiltration (NF), and RO have been adopted in the treatment of wastewater to improve the treatment efficiency and treated water quality (Nataraj et al. 2006; Prodanović and Vasić 2013). Among them, NF is the most attractive method for further treatment of anaerobiclly and aerobically treated distillery wastewater to achieve maximum recovery and reuse of distillery wastewater, which offers the advantages of less energy consumption and high efficiency of separation in most of the operations (Rai et al. 2008). The NF membrane was developed with the necessity of lowering the pressure of RO process with reasonable permeate flux and good separation, particularly of organic components. NF membranes are usually charged with a selective layer of thickness of ~1 μm coated over the ultraporous membrane that controls all the rejection and flow properties by diffusion, convection, and Donnan exclusion mechanisms. It can be either considered as an interaction process step for intermediate separation of solutes or used as an end-of-pipe treatment for reuse of the treated water. The space requirements are less and there is no generation of sludge. There is a reduction in the freshwater usage as the water can be completely recycled and reused. But the high cost of membranes and equipment, the lowered productivity with time due to fouling of the membrane, and the disposal of concentrates are the limitations. Melanoidins are effectively decolorized using chemical oxidizing agents and seem to hold potential for future use in the distillery industry. RO and MF have also been advocated and practiced in distillery effluent treatment. The technique generates about 50% colorless reusable water and the balance 50% concentrate that can be easily composted by available press mud. The concentrated stream contains almost double the levels of COD, BOD, and TDS. Rodriguez-caballero et al. (2012) treated high ethanol concentration wastewater by biological sand filters. Furthermore, they analyzed the bacterial community structure by means of molecular fingerprinting technique (DGGE) from the BSF sediment samples. DGGE showed that amendment with high concentrations of ethanol destabilized the microbial community structure, but that nutrient supplementation countered this effect.

5.3.10 Advanced Oxidation Processes

AOPs are considered highly effective wastewater treatment technologies for removing low biodegradability, refractory, inhibitory, or high chemical stability pollutants. The mechanisms of AOP rely on the in situ formation of highly reactive oxidant species, mainly hydroxyl radicals ($\cdot OH$), highly effective, powerful, and ubiquitous in nature and non-selective with electrophilic behavior and have a redox potential of 2.8V, which accelerate the oxidation and degradation of a wide range of contaminants in wastewater by abstracting a hydrogen atom from aliphatic carbon, or adding a hydrogen atom to the double bonds and aromatic rings (Deng and Zhao 2015; Kumar et al. 2021a). The $\cdot OH$ in compounds destroy and even mineralize them into CO_2, H_2O, and inorganic salts to some extent. Other radicals and active oxygen species generated in AOPs are superoxide radicals $\cdot O_2^-$, hydroperoxyl radicals $\cdot HO_2^-$, sulfate radicals ($\cdot SO_4^-$), and organic peroxyl radicals ($\cdot ROO$). The generation of these reactive agents can be achieved by means of several processes, including sonolysis, ozone-based processes (ozonation), UV irradiation, Fenton oxidation, photo-Fenton oxidation, ultrasound, photocatalysis, and various combinations of these technologies.

5.3.10.1 Fenton Oxidation

Fenton oxidation is a very effective technology to treat distillery wastewater. In addition to improving the biodegradability of the effluent, the Fenton process removes the remnant COD, toxicity, and color of the wastewater. In the Fenton treatment process, a mixture of H_2O_2 and ferrous iron salts (Fe_2^+) generates highly $\cdot OH$ radical, which involves a complex reaction sequence and ultimately leads to organic removal of the wastewater. The Fe_2^+ initiates and catalyzes the decomposition of H_2O_2 in acidic conditions, resulting in the generation of $\cdot OH$ (Sievers, 2011; Boczkaj and Fernandes, 2017). The Fenton process mechanism is presented by the following reactions:

$$Fe^{2+} + H_2O_2 \rightarrow Fe_3^+ + OH^- + \cdot OH \tag{5.1}$$

$$Fe^{2+} + \cdot OH \rightarrow Fe^{3+} + OH^- \tag{5.2}$$

$$\cdot OH + RH \rightarrow H_2O + R \cdot \tag{5.3}$$

$$R \cdot + Fe^{3+} \rightarrow R^+ + Fe^{2+} \tag{5.4}$$

There are also reports of enhanced performance in COD removal, an increase in biodegradability, and the detoxification of complex wastewater after treatment with photo-Fenton oxidation (Derakhshan and Fazeli 2018). The toxicity of H_2O_2 to several microorganisms and the use of excess amounts of H_2O_2 could possibly deteriorate the overall degradation efficiency for cases where the Fenton process is followed by biological oxidation. Modified natural zeolite as heterogeneous Fenton catalyst in treatment of recalcitrant industrial effluent has been reported by Arimi (2017). The sulfuric acid-ferrous-modified catalysts showed the highest affectivity, which achieved 90% color and 60% TOC (total organic carbon) removal at 150 g/L pellet catalyst dosage, 2 g/L H_2O_2, and 25 °C. The heterogeneous Fenton with the same catalyst caused improvement in the biodegradability of anaerobic effluent from 0.07 to 0.55. The catalyst was also applied to pretreat the raw molasses distillery wastewater and increase its biodegradability by 4%. The color of the resultant anaerobic effluent was also reduced. Arimi et al. (2015) developed a polishing step after anaerobic digestion for the colorant elimination from melanoidin-rich wastewater using natural manganese oxides. It was observed that the kinetics of colorant elimination was best described by the second-order equation, with a significant dependence on pH. Melanoidins have many amine groups in their structure, which indicate the possibility of the oxidation and decoloration by manganese oxides. Hayase et al. (1984) investigated the decolorization and decomposition products of melanoidins prepared from a glucose-glycine reaction system by using H_2O_2 under neutral (pH 7) and alkaline (pH 10) conditions at 37 °C for 28 h. Melanoidins were decolorized to 64%–97%, respectively, at pH 7 and 10. Generally, H_2O_2 reacts with the hydroxyl anion to give mainly perhydroxy anion (HOO-), which has a strong nucleophilic activity. The perhydroxy anion is considered to attack nucleophilically the carbonyl groups in melanoidins. Decolorization and degradation of melanoidins occur markedly in the alkaline medium due to the abundant presence of the perhydroxyl anion. The presence of metals (Fe, Cu, etc.) can induce an autodegradation of H_2O_2 to form the free radicals such as OH, OOH, and OO⁻, which may positively react with melanoidins. It is, however, well-known that melanoidin itself is capable of producing reductones and forming chelates with metallic ions. Consequently, it is suggested that very small amounts of metallic ions existing in the present reaction system interact with melanoidins and radical reactions due to H_2O_2 may be suppressed by the formation of the melanoidin-metal complex. The major components in the ether-soluble fraction obtained from melanoidins by oxidative degradation of alkaline H_2O_2 were identified as 2-methyl-2,4-pentanediol, *N,N*-dimethylacetamide, phenol, acetic acid, oxalic acid, methylpropane dioic acid, propanedioic acid, 2-furancarboxylic acid, butanedioic acid, 2-hydroxypropanoic acid, 2,5-furandicarboxylic acid, and 5-(hydroxymethyl)-2-furancarboxylic acid by GC-MS analysis (Hayase et al. 1984)

5.3.10.2 Wet Air Oxidation

Wet air oxidation (WAO), also called thermal liquid-phase oxidation, is useful for treatment of effluents containing high content of organic matter or toxic contaminants. It is carried out at elevated temperature (125–320 °C) and pressure (0.5–20 MPa) using a gaseous source of oxygen (O_2)/air in which there is generation of active oxygen species such as hydroxyl radicals. At elevated temperatures and pressures, the solubility of O_2 in the aqueous solution increases and provides a strong driving force for oxidation of pollutants. Generally, elevated pressures are maintained to keep aqueous solutions in a liquid state. A few studies also explored the possibilities of complex wastewater treatment by the WAO, which can be followed as a pretreatment step to enhance the biodegradability and facilitate biogas generation in subsequent biological treatment processes (Padoley et al. 2012a,b; Goto et al. 1998). A lab-scale WAO reactor with biomethanated distillery wastewater (COD: 40,000 mg/L; initial biodegradability index (BI) of 0.17) was used to demonstrate the proof of concept (Padoley et al. 2012a,b). WAO-induced enhanced biodegradability of distillery effluent was reported by Malik et al. (2014). WAO pretreatment of bio-methanated distillery effluent resulted in substantial enhancement in the BI (up to 0.8). WAO pretreated effluent on anaerobic digestion indicated favorable biogas generation with methane content up to 64% along with concomitant COD reduction up to 54.75%. The HPLC analysis indicated that the pretreatment facilitated degradation of major color-containing compounds, namely, melanoidins, up to 97.8% (Sarat Chandra et al., 2014). A hybrid pretreatment technique was explored for the selective improvement in biodegradability of biomethanated spent wash (Bhoite and Vaidya 2018a,b). The introduction of catalyst in WAO, called catalytic wet air oxidation (CWAO), facilitates the oxidation process at considerably low temperature and pressure. CWAO is useful for treatment of effluents where the concentration of organic contaminants is too low for the incineration process to be economical. Further, the CWAO leads to formation of carbon dioxide and water and there is no formation of by-products. Ozone is a more power-ful oxidant than chlorine and other compounds (hydrogen peroxide, potassium permanganate, chlorine dioxide, and bromine). In another study, the iron-catalyzed WAO of biomethanated spent wash exhibited high BOD_5 (8100 mg/L) and COD (40,000 mg/L) for enhanced biogas recovery, as reported by Bhoite and Vaidya (2018a). The severe operating conditions result in high installation cost (material of construc-tion of equipment to withstand high temperature and pressure) and operating cost, which are the major obstacle for this technology.

5.3.10.3 Photodegradation

Photocatalytic degradation or photocatalysis is an attractive, efficient, and cost-effective treatment technology to enhance the biodegradability of hazardous and non-biodegradable contaminants, such as persistent organic pollutant (Takle et al. 2018). Photocatalytic treatment process is based on the com-bination of oxidizing agents with an appropriated catalyst and/or light. David et al. (2015) studied deg-radation of organic pollutants in the form of color using nano-photocatalyst prepared using aluminum oxide (Al_2O_3) nanoparticle and kaolin clay. The nano-Al_2O_3/kaolin composite photocatalyst prepared by calcination at high temperature possessed and demonstrated good absorption of visible light range, which can degrade the intense color of the distillery spent wash effluent. The key process parameters such as catalyst dosage, pH, temperature, and agitation speed were optimized by Taguchi method and a maximum of 80% decolorization of distillery spent wash was achieved. Photodegradation of spent wash and Jakofix dye using vanadium-doped titanium dioxide (V-TiO_2) nanoparticles prepared using a simple sol-gel method based on aqueous titanium peroxide was studied by Takle et al. (2018). The degradation of colored compounds in spent wash was monitored by gel permeation chromatography (GPC), which showed the degradation of high-molecular-weight compounds into low-molecular-weight fractions. The catalyst decomposed 90% of Jakofix red dye (HE 8BN) in 3.5 h and 65% of spent wash in 5 h under irradiation with natural sunlight, whereas Degussa P-25 TiO_2 was only able to decom-pose 35% of the dye and 20% of spent wash under identical reaction conditions. A cycling stability test showed the high stability and reusability of the photocatalyst for degradation reactions, with a recovery of around 94–96%. The V-TiO_2 powder photocatalyst exhibited very high activity in the photocatalytic degradation of spent wash and Jakofix red dye under natural sunlight. Among the photocatalysts that were investigated, 1% V-TiO_2 was found to be the most active for the degradation of color in spent wash (65%) and Jakofix red dye (90%). The photocatalyst was tested for its activity and exhibited higher activity under natural sunlight than under an artificial source of energy (Xe lamp). A study on the syn-ergy of a combined AD (anaerobic digestion) and UV photodegradation treatment of distillery effluent was carried out in fluidized bed reactors to evaluate pollution reduction and energy utilization efficien-cies (Apollo and Aoyi (2016). The combined process improved color removal from 41% to 85% com-pared to that of AD employed as a stand-alone process. An overall corresponding total organic carbon (TOC) reduction of 83% was achieved. The bioenergy production by the AD step was 14.2 kJ/g TOC biodegraded, while UV lamp energy consumption was 0.9 kJ/mg TOC, corresponding to up to 100% color removal. It was concluded that a combined AD-UV system for treatment of distillery effluent is effective in organic load removal and can be operated at a reduced cost. There is a possible synergy when the two processes are integrated, with AD preceding UV, where AD removes the high COD/TOC and UV removes the recalcitrant color at a reduced cost. Mabuza et al. (2017) investigated the synergy of integrated anaerobic digestion and photodegradation using hybrid photocatalyst for molasses wastewater treatment. In this study, a hybrid photocatalyst consisting of titanium dioxide (TiO_2) and zinc oxide (ZnO) was used for photocatalytic degradation of effluent. Biodegradation at thermophilic conditions in the bioreactor achieved high TOC and COD reductions of 80 and 90%, respectively, but with an increased color intensity. Contrastingly, UV photodegradation achieved a high color reduction of 92% with an insignificant 6% TOC reduction, after 30 min of irradiation. During photodegradation, the mineralization of the biorecalcitrant organic compounds led to the color disappearance.

5.3.10.4 Ozonolysis

Ozone has many properties desirable for the treatment of the wastewater. First, it is a powerful oxidant capable of oxidative degradation of many organic compounds, including taste, odor, color, nitro-aromatics, phenols, azo dyes, chlorobenzene, and particle removal, and also results in oxidation products that are more biodegradable. Ozone is a very reactive, unstable gas and has low solubility in water. Ozone upon dissolution reacts with the organics/inorganics present in the wastewater either directly at acidic/ neutral conditions and/or indirectly by decomposition at highly alkaline pH. Ozone also results in the formation of highly reactive hydroxyl radical in the system, which has higher oxidation potential as compared to ozone itself. The oxidation potential of ozone is 2.07 as compared to 1.36 for chlorine and 1.78 for hydrogen peroxide. Kim et al. (1985) examined the decolorization and degradation products of melanoidins prepared from a glucose-glycine system after ozonolysis. Melanoidins were decolorized to degrees of 84% and 97% after ozonolysis at −1 °C for 10 min and 90 min, respectively, and the mean molecular weight of melanoidins decreased from 7000 to 3000 after ozonolysis for 40 min. Melanoidins after ozonolysis were fractionated into acidic, neutral, and basic fractions with ether. The major degradation compounds in the acidic fraction were identified as methoxyacetic acid, 2-hydroxybutanoic acid, glycolic acid, propanedioic acid, and butanedioic acid. Among the minor products, compounds having methyl side chains, such as 3-hydroxy-3-methylbutanoic acid and 2-hydroxy-2-methylpropanoic acid, were also identified. Ozone attacks organic compounds electrophilically and reacts with H_2O in aqueous solution to form H_2O_2 in part. As this experiment was performed on the acidic side in the pH range of 3–4, the nucleophilic attack of H_2O_2 secondarily formed on the reaction of ozone with H_2O was considered to be suppressed much more than on the alkaline side. Consequently, the electrophilic reaction on ozone is proposed to mainly occur in the present experimental system. Especially, ozone as an amphoteric ion makes an electrophilic attack in an electron-rich system such as a carbon-carbon double bond and causes cleavage of that bond. Melanoidins have C=C double bonds in their molecules. Kim et al. (1985) stated that the decolorization and the decrease in the molecular weight of melanoidins are due to cleavage of C=C double bonds in their molecules by ozonolysis (Kim et al. 1985). Malik et al. (2019) evaluated the effect of ozone pretreatment on the rate of composting process and the quality of compost obtained. They recommended that the ozonation-based pretreatment of complex biomethanated distillery wastewater was effective in enhancing its biodegradability along with color reduction, thereby facilitating enhanced composting. They showed that the carbons adjacent to the oxygen or nitrogen atom were not affected by the ozone treatment. These saturated and aliphatic carbons are supposed to form the principal skeleton or backbone of melanoidins. Moreover, the ozone treatment is supposed to lead to cleavage of the C=C or C=N bonds. These unsaturated bonds have been suggested to be important for the structure of the chromophore. Studies using glycine-2-^{13}C have revealed that most of the glycine was incorporated into melanoidins, and the authors speculated that the nitrogen in melanoidins was mainly due to conjugated enamine linkage. Ozonolysis process has two main drawbacks. The first is low oxidation efficiency due to the selective nature of ozone, limited mass transfer, and slow rate of reaction. The second is related to the high energy consumption for ozone generation that makes it costly. Hence, the technologies are required to increase the efficiency of the ozonation process at low cost with high efficiency. Catalytic/hybrid ozonation such as (O_3+OH), UV (O_3+UV), peroxone (O_3+H_2O_2) is a promising technology for the wastewater treatment as it can reduce operating cost and reaction time to achieve the same degradation efficiency with increased rate of degradation and mineralization of recalcitrant organic compounds present in the wastewater. It is an environment-friendly technique, but it is not effective on all dyes as its oxidation potential is not very high. Further, the process needs to be activated by some other means like UV light, inorganic salts, ozone, or ultrasound/sunlight.

5.4 Biological Treatment

Biological-based technique is emerging as a nature-friendly, permanent and greater public acceptance, less expensive, and minimal site disruption process of waste elimination of any wastewater treatment plant that treats wastewater from either municipality or industry having soluble organic impurities or a mix of the two types of wastewater sources. A typical COD/BOD ratio of 1.8–1.9 indicates the suitability of the effluent for biological treatment. Biological processes have utilized various microbes such as algae, fungi, yeast, and bacteria and plants or their enzymes for effective decolorization of distillery effluent. In recent years, biological processes have attracted the attention of researchers/scientists and helped the development of an efficient eco-friendly effluent treatment system.

5.4.1 Microbial Treatment

Microorganisms constitute the oldest members of living systems and possess higher adaptability to thrive in adverse conditions. The microbial community plays an important role in the soil system, especially in extreme ecosystems where they are responsible for the majority of organic matter decomposition (Chandra and Kumar 2017b; Kumar 2018a; Kumar et al. 2020a,b). Moreover, the use of microbes for degradation and detoxification of pollutants is now being increasingly applied as the technology of choice for cleanup or restores contaminated sites to environmentally sustainable (Chandra and Kumar 2015a; Kumar et al. 2018; Kumar and Chandra 2020a). Distillery wastewater treatment using environment-friendly, non-pathogenic, and ecologically sustainable microorganisms put an additional insight over existing techniques of microbial bioremediation. Decolorization of melanoidins and/or melanoidins-containing distillery effluent by using a number of microorganisms has been reported by various researchers (Chowdhary et al. 2020).

5.4.1.1 Algal-Assisted Treatment

Photosynthetic organisms, like algae and cyanobacteria, are ubiquitous in nature and have been studied for their decolorization potential of wastewater and it has been proposed that algae and cyanobacteria are proficient in degradation of melanoidins-containing distillery effluent (K. Sankaran et al., 2014). The remediation by microalgae technology (phycoremediation) is an environment-friendly approach that offers elegant solution for the tertiary wastewater treatment due to its ability to utilize the pollutants for its growth with no generation of secondary pollution (Krishnamoorthy et al. 2017; Amenorfenyo et al. 2020). Microalgae such as *Chlorella sorokiniana* (Solovchenko et al. 2014), *Spirulina* sp. (Krishnamoorthy et al. 2019), *Oscillatoria* sp. (Krishnamoorthy et al. 2017), and *Oscillatoria boryana* (Francisca Kalavathi et al. 2001) are useful in treating the distillery wastewater and the efficiency of treatment is promising. Microalgae can exhibit mixotrophic or heterotrophic mechanism while treating the wastewater. It can assimilate organic pollutants into its cellular constituents such as carbohydrates, proteins, lipids, and inorganic pollutants as nutrients for its growth and multiplication (Amenorfenyo et al. 2020). The anaerobically digested distillery wastewater (ADDW) provides satisfactory conditions for the growth of microalgae. The cultivation of algae on ADDW could replace the water and nutrient required for its survival and proliferation (Chinnasamy et al., 2010). The microalgal biomass obtained after the treatment could be used as a raw material for many applications such as biofuel (biodiesel/biogas), feed for aquaculture, animal feed, fertilizer in agriculture, and a source of high-value chemicals (pigments). However, it requires a quality check with regard to the cell wall thickness, C/N ratio, and lipid proportion for deciding the route of algae biomass utilization by economic analysis. The *Spirulina* sp. showed the greater adaptability in the distillery environment. The species were sustainable in the 500-L capacity photobioreactor (Krishnamoorthy et al., 2019). It has significantly reduced the pollution load of distillery wastewater up to 60–70% by the uptake of inorganic compounds as its nutrients. Further, the growth of algae was supported by the customized photobioreactor and the system had its own advantages such as less space requirement, moderate energy requirement (292 W/m^3), and control over insect/contamination entries through the presence of submerged mixers. Francisca Kalavathi et al. (2001) reported the degradation and metabolization of melanoidin pigment in distillery effluent by the marine cyanobacterium *Oscillatoria boryana* BDU 92181. *O. boryana* BDU 92181 used the melanoidin as nitrogen and carbon source leading to decolorization. The organism decolorized pure melanoidin pigment (0.1% w/v) by about 75% and crude pigment in the distillery effluent (5% v/v) by about 60% in 30 days. The mechanism of color removal is postulated to be due to the production of H_2O_2, hydroxyl anions (HOO-), and molecular oxygen, released by the *O. boryana* BDU 92181 during photosynthesis. H_2O_2 generated in cyanobacteria during photosynthesis can react with hydroxyl anion to give mainly HOO-, which has a strong nucleophilic activity and can also help decolorization of melanoidin. Color reduction can also be brought about by active oxygen, released during the photolysis of water by the cyanobacterium. Melanoidin can be removed by algal/cyanobacteria through three diverse mechanisms of assimilative chromophores usage: (i) for the creation of algal/cyanobacterium biomass, (ii) H_2O as well as CO_2 alteration from colored to non-colored molecules, (iii) chromophores adsorption on algal/cyanobacteria biomass. A study demonstrated the feasibility of combining microalgae and macrophytes for bioremediation of recalcitrant industrial wastewater. Ravikumar et al. (2013) studied the response surface methodology and artificial neural network for modeling and optimization of distillery spent wash treatment using *Phormidium valderianum* BDU 140441. The effect of initial pH (6–10), temperature (24–32 °C), and light intensity (20–54 W/m^2) was studied using single factorial design and achieved a maximum decolorization of 74.5% with COD reduction of 83.48%. A 23 full factorial experimental central composite design (CCD) of response surface methodology (RSM) was used to investigate the interaction effect between these variables and the optimal values. The predicted results showed that a maximum decolorization of 85.5% and COD reduction of 87.29% was achieved under the optimum conditions of 8 pH, 30 °C, and light intensity of 36 W/m^2. Valderrama et al. (2020) performed laboratory-scale experiments to develop a procedure for biological treatment of recalcitrant anaerobic industrial effluent (from ethanol and citric acid production) using first the microalga *Chlorella vulgaris* followed by the macrophyte *Lemna minuscula*. The wastewater was diluted to 10% of the original concentration with wash water from the production line. Within 4 days of incubation in the wastewater, *C. vulgaris* population reduced ammonium ion (71.6%), phosphorus (28%), and COD (61%), and dissolved a floating microbial biofilm after 5 days of incubation. Consequently, *L. minuscule* was able to grow in the treated wastewater (from 7 to 14 g/bioreactor after 6 days), precipitated the microalgal cells, and reduced other organic matter and color (up to 52%) after an additional 6 days of incubation. However, *L. minuscula* did not improve removal of nutrients. Solovchenko et al. (2014) demonstrated the possibility of efficient phycoremediation of anaerobically digested alcohol distillery wastewater with the use of the high-density semibatch culture of *C. sorokiniana* strain cultivated in a photobioreactor monitored online via chlorophyll fluorescence. Nearly complete deodoration and removal of the bulk of inorganic nutrients and organic matter from the distillery wastewater were achieved within 3–4 days. Growth of microalgae is a photosynthetic process that uses sunlight to synthesize the complicated carbon/inorganic molecule present in the effluent (Sankaran and Premalatha 2018). The advantages of phycoremediation are as follows: (i) The nutrients are available in the effluent itself; (ii) no further water is added for the growth of microalgae; (iii) sufficient light energy required for photosynthesis is available in tropical countries; and (iv) the biomass yield can be further used for production of biogas or biodiesel or fertilizer. Algae utilized the inorganic compounds from the wastewater as its nutrients and thus resulted in the substantial removal of pollutants from the wastewater. The possibilities of the effect of *Spirulina* sp. biomass (ranging from 1 to 5 g/L) value on the treatment of anaerobically digested distillery effluent was carried out under xenon lamp illumination of 10,000 ± 200 lx with the photo-period of 16:8 (Sankaran and Premalatha 2018). The reduction of COD up to 15–23%, total carbon (TC) up to 18–31%, total inorganic carbon (TIC) up to 31–51%, and TOC up to 1–8% were observed for the algae concentration ranging from 1 to 5 g/L over a period of 5 days. The reduction of carbonate up to 32–56%, ammonium 48–72%, phosphate 18–100%, chloride 1–8%, potassium 2–7%, magnesium 10–23%, and sodium 1–9% were observed by ion chromatography analysis in the wastewater after treating with algal biomass ranging from 1 to 5 g/L. The calorific value of dry algae biomass was found as 12 MJ/kg, which might be utilized for the production of biofuel. Similarly, a laboratory and outdoor study was also conducted by Krishnamoorthy et al. (2019) to evaluate the treatability of anaerobically digested distillery wastewater in a customized photobioreactor using blue-green microalgae *Spirulina* sp. The phycoremediation has shown significant pollution reduction in the wastewater. In this study, the treatability of wastewater of COD 30,000–40,000 ppm was carried out in those photobioreactors with the *Spirulina* sp. of culture volume fraction 0.8 and 0.93 under xenon lamp and sunlight, respectively. The COD and TDS reduction were 60–70% in both the volume fractions of the culture. Ion chromatography analysis indicated the reduction of major inorganic pollutants in the wastewater by the *Spirulina* sp. Algae that play a major role in remediation of organic contaminants and nutrients and used as pollution indicators for various toxicants in wastewater needs further research.

5.4.1.2 Bacteria-Assisted Treatment

Among many methodologies used for effluent treatment (economically viable and widely adopted), the bioremediation using bacteria is well-known (Kumar 2018b; Kumar et al. 2018; 2020a). Bacteria are important contributors to the transformation of complex organic, inorganic, and organometallic compounds in wastewater treatment systems and are essential for the optimal operation and preservation of biological treatment systems (Chandra and Kumar 2015a; Kumar and Chandra 2018a; 2020a). The main advantage to work with bacteria is that they are easy to culture and grow more quickly as compared to other microbes. They are able to catabolize phenolics and chlorinated and aromatic hydrocarbon-based organic pollutants, which can be decomposed by using them as an energy source and have the ability to transform and detoxify melanoidins. Many studies have shown favorable results in identifying bacteria that can degrade melanoidins and Maillard reaction products containing distillery effluent at a faster rate (Murata et al. 1992; Bharagava et al. 2009; Yadav and Chandra 2012). Different bacterial group under traditional aerobic, facultative anaerobic, and extreme oxygen-deficient conditions cause melanoidins reduction for decolorization. The chemical reaction during the reduction of the azo dyes starts with the breaking of double bonds (-C=N- and C=C-) under aerobic environment by the ligninolytic enzyme that results in a colorless solution of containing potentially hazardous compounds (Kumar and Chandra 2018a; Chandra et al. (2018a). Aerobic bacteria are widely used for melanoidins decolorization due to their high activity, extensive distribution, and strong adaptability. Melanoidin containing distillery effluent has been decolorized by bacteria, including *Paracoccus pantotrophus* (Santal et al. 2016); *Pseudomonas aeruginosa* (David et al. 2015); *Bacillus* sp. (Kaushik and Thakur 2009); *Klebsiella oxytoca, Serratia mercescens, Citrobacter* sp. (Jiranuntipon et al. 2008); *P. aeruginosa* PAO1, *Stenotrophomonas maltophila*, and *Proteus mirabilis* (Mohana et al., 2007); *Klebsiella pneumoniae, Salmonella enterica, Enterobacter aerogenes*, and *Enterobacter cloaceae* (Kumar and Chandra 2018a); *Lactobacillus kefiri* (Omar et al. 2020); *Pseudomonas* sp. (Chavan et al. 2006); and *Exiguobacterium acetylicum* strain QD-3, *Bacillus cereus* strain H3, *E. cloaceae* strain CPO, and *Enterobacter* sp. LY402 (Ruhi et al. 2017).

5.4.1.2.1 Using Pure Bacterial Culture

Pure bacterial cultures have been studied in order to develop bioprocess for melanoidins decolorization in molasses wastewater. The decolorization of spent wash and pollutants utilizing *Bacillus* sp., isolated by distillery mill waste polluted soil, has been studied by Kaushik and Thakur (2009) in static condition. The investigators achieved maximum decolorization of distillery spent wash by the *Bacillus* sp. Reduction in color (85%) and COD (90%) was observed within 12 h after optimization by the Taguchi method. Sankaran et al. (2015) reported the bacteria-assisted treatment of anaerobically digested distillery wastewater as the high COD (30,000–35,000 ppm), BOD (8000–10,000 ppm), total solids (60,000–65,000 ppm), and presence of organic compounds even after anaerobic treatment of wastewater. Bacterial culture rich in *Pseudomonas* sp. was effective in the treatment of distillery effluent with a COD 25,000–32,000 ppm pollution load. In the presence of excess of carbohydrate source, which is a physiological stress, several bacterial strains produce polyhydroxybutyrate (PHB), an intracellular polymer, which is synthesized, is primarily a product of carbon assimilation, and is employed by microorganisms as an energy storage molecule to be metabolized when other common energy sources are limitedly available. Efforts have been made by Davis and his team in 2015 to check whether the PHB has any positive effect on spent wash decolorization. When a combination of PHB and the isolated bacterial culture of *P. aeruginosa* was added to spent wash, a maximum color removal of 92.77% was found, which was comparatively higher than the color removed when the spent wash was treated individually with the PHB and *P. aeruginosa*. PHB behaved as a support material for the bacteria to bind to it and thus develops biofilm, which is one of the natural physiological growth forms of microorganisms and resulted spent wash decolorization. Under aerobic and optimized environmental conditions (pH 7.5 and temperature 37 °C), the decolorization of melanoidin via *Alcaligenes faecalis* SAG_5 with decolorization efficiency of 72.6% at fifth day of cultivation has been reported (Santal et al. 2011). A novel thermotolerant bacterium *Pediococcus acidilactici* B-25 exhibited maximum 79% decolorization, 85% COD, and 94% BOD reduction from distillery effluent at 45 °C using 0.1% glucose, 0.1% peptone, 0.05% $MgSO_4$, 0.05% K_2HPO_4 at pH 6.0 within 24 h under static condition (Tiwari et al. 2012). Kaushik and Thakur (2014) investigated the production of laccase and optimization of its production by *Bacillus* sp. using distillery spent wash as inducer. The usefulness of the Taguchi method for optimization of culture conditions was investigated with five selected factors at four levels, and it was observed that the optimized medium resulted in a ninefold increase in extracellular laccase production compared with the control. The optimized medium composition for laccase production was dextrose (1%), tryptone (0.1%), $CuSO_4$ (1 mM), and an inducer (distillery effluent 10% [v/v]) at pH 7, which altogether resulted in 107.32 U/mL extracellular laccase activity. Hence, the Taguchi approach proved to be a reliable tool in optimizing culture conditions and achieving the best possible combination for enhanced laccase production. Thiyagu and Sivarajan (2018) isolated and characterized a novel bacterial strain *Stenotrophomonas* sp. from lab-scale hybrid UASB reactor treating distillery spent wash. The hybrid UASB reactor was operated for 360 days with 24 h HRT and has an optimum COD removal efficiency of 83.87% at an organic loading rates (OLRs) ranging 0.25–27.40 kg COD/m^3per day. A highly effective decolorizing bacterium strain, *P. pantotrophus* SAG_1, decolorized melanoidins up to 81.2 ± 2.43% in the presence of glucose and NH_4NO_3 (Santal et al., 2016). Melanoidin degradation in connection of effluent decolorization with pure cultures of bacterial strains is effective; however, distinct bacterial isolates frequently could not be able to decolorize melanoidins entirely and are also responsible for the production of toxic compounds as the transitional products, which need further degradation (Santal et al. 2011). Therefore, the treatment schemes usually consist of mixed microbial communities that attain an advanced level of mineralization and biodegradation majorly because of the synergistic metabolic actions of the microbial communities; these are also significantly beneficial as compared with the use of pure cultures. **Table 5.2** provides a list of some potential bacterial strains that have been reported in degradation and decolorization of distillery effluent under controlled conditions.

5.4.1.3 Utilizing Co-Culture and Mixed Bacterial Cultures

The co-cultures, or hybrid bacterial cultures, can lead to a higher level of biodegradation, especially because melanoidins molecules are attacked in different positions. Similarly, the consortium between *K. pneumoniae*, *S. enterica*, *E. aerogenes*, and *E. cloaceae* treating a synthetic as well as natural melanoidins achieves better results of biodegradation (Kumar and Chandra 2018a; Chandra et al. 2018a). This is possible due to the action of manganese peroxidase and laccases activity, occurring in *K. pneumoniae*, *S. enterica*, *E. aerogenes*, and *E. cloaceae* for the biodegradation of melanoidins. Bhargava and Chandra (2010) determined that the bacterial consortium consisting of *Bacillus licheniformis*, *Bacillus* sp., and *Alcaligenes* sp. could decolorize major color-containing compounds of distillery wastewater under aerobic conditions. The authors' findings revealed that the rigorous action of isolates may substantially diminish the 70% color of wastewater in the presence of glucose (1.0%) and peptone (0.1%) at pH 7.0 and temperature 37 °C. Further, Yadav and Chandra (2012) reported that a developed bacterial consortium comprising *P. mirabilis*, *Bacillus* sp., *Raoultella planticola*, and *Enterobacter sakazakii* could eliminate the organic compounds of molasses melanoidin from biomethanated distillery spent wash (BMDS) in the ratio of 4:3:2:1

TABLE 5.2

Decolorization of Distillery Effluent and Reduction of Pollution Parameters by Pure Bacterial Strains as Reported by Different Group of Researchers

Bacterial strain	Wastewater	Abs.	BOD (%)	COD (%)	Decol (%)	Incub	pH	Temperature (°C)	References
Lactobacillus plantarum MiLAB393	Sugar beet molasses vinasse	–	–	–	24.1	–	–	35.8	Wilk and Krzywonos (2020)
Lactobacillus plantarum, L. casei, and *Pediococcus parvulus*	Sugar beet molasses vinasse	–	–	–	28.36	72 h	–	35.8	Wilk et al. (2019)
Pseudomonas aeruginosa	Distillery spent wash	–	–	–	92.77	–	–	–	David et al. (2015)
Alcaligenes faecalis SAG$_5$	Distillery effluent	–	–	–	72.6	5 d	–	37	Santal et al. (2011)
Lactobacillus plantarum No. PV71-1861	Molasses wastewater	–	–	–	68.12	7 d	–	30	Tondee and Sirianuntapiboon (2008)
Lactobacillus plantarum MiLAB393 Marta	Distillery wastewater	–	–	–	30 (v/v)	–	–	35.8	Wilk and Krzywonos (2020)
Pseudomonas sp.	Distillery spent wash	–	–	–	56	72 h	–	35	Chavan et al. (2006)
Acetogenic bacteria strain No. BP103	Molasses wastewater	–	–	–	76.4	5 days	–	35	Sirianuntapiboon et al. (2004)
Bacillus sp.	Distillery spent wash	475	–	90	85	12 h	–	35	Kaushik and Thakur (2009)
Pseudomonas sp.	Distillery spent wash	475	–	63	56	72 h	6.8–7.2	30–35	Chavan et al. (2006)
Pediococcus acidilactici	Distillery spent wash	–	94	85	79	–	–	45	Tiwari et al. (2012)

within 10 days under optimized static conditions. Bacterial consortium showed manganese per-oxidase and laccase activity during melanoidins decolorization. The potential manganese peroxidase (MnP) producing bacterial consortium of *Bacillus* sp., *R. planticola*, and *E. sakazakii* showed maximum decolorization (60%) of sucrose-aspartic acid Maillard product (SAA-MP) (2400 mg/L) in modified GPYM medium at optimized nutrient, pH (7.0), shaking speed (180 rpm), and temperature (35 °C) after 144-h incubation. The addition of D-xylose enhanced the decolorization of SAA-MP from 60% to 75% along with reduction of BOD and COD (Yadav et al. 2011). Adikane et al. (2006) achieved maximum decolorization (81%) of molasses spent wash (MSW; 12.5%) after 18 days of incubation in the absence of any additional carbon or nitrogen source using soil as inoculum (10%). Omar and his team (2020) isolated potential manganese peroxidase producing bacteria two *Klebsiella* sp. (B2–B3) and *Escherichia coli* (B4) and one strain of *L. kefiri* (B1) from a bioreactor treating a mixture of municipal and molasses wastewater that showed high synthetic melanoidin tolerance. These isolated strains reduced the COD by more than 50% or more than 70% of the initial value with or without additional supplementation of the reaction mixture, respectively. All tested isolates showed increased ability for reducing the organic matter content of a raw melanoidin solution exceeding 65%, favoring therefore the utilization potential of the isolated strains for the biological processing of molasses wastewaters. Ghosh et al. (2004) isolated and identified six bacterial strains *Pseudomonas*, *Enterobacter*, *Stenotrophomonas*, *Aeromonas*, *Acinetobacter*, and *Klebsiella* capable of using recalcitrant compounds of molasses spent wash as sole carbon source from the soils of abandoned sites of distillery effluent discharge and characterize their ability of reducing the COD of the spent wash. The extent of COD (44%) reduced collectively by the six strains was equal to that reduced individually by *Aeromonas*, *Acinetobacter*, *Pseudomonas*, and *Enterobacter*. With spent wash as sole carbon source, the COD reducing strains grew faster at 37 °C than 30 °C. Chandra and Kumar (2017a) characterized indigenous *Bacillus* and *Stenotrophomonas* species grown in an organic acid and EDC-rich environment of distillery spent wash. In addition, scientists have detected and characterized various recalcitrant organic compounds discharged in spent wash after ethanol production. These findings indicated that these autochthonous bacterial communities were pioneer taxa for in situ remediation of this hazardous waste during ecological succession. Chowdhary et al. (2020) treated distillery effluent containing several toxic organic compounds with the mixed culture of *Staphylococcus saprophyticus* and *Alcaligenaceae* sp. under aerobic conditions. Authors stated that mixed bacterial culture exhibiting decolorization ability maximally remove 71.83% of distillery wastewater. The mixed bacterial communities and/or bacterial consortium used by the researchers in degradation and decolorization of distillery effluent are listed in **Tables 5.3 and 5.4**.

5.4.1.4 Fungal-Assisted Treatment

Being an economical, eco-friendly, less-sludge-producing process, fungal treatment is very advantageous process used by researchers to find a new way for remediation. In contrast, the bacterial decolorization is usually faster, but it may require a mixed community to decolorize melanoidins through combined metabolic mode of individual culture. Fungi are also able to grow in a wide pH range at levels below ideal for most common soil bacteria. Fungi can grow at lower levels of moisture than bacteria. Contrary to bacteria, fungi robustly degrade multifaceted organic complexes by utilizing methods such as production of extracellular ligninolytic enzymes comprising manganese peroxidase (MnP), laccase (Lac), and lignin peroxidase (LiP); thus, scientists are more interested in using fungi in the current scenario. In the environment, fungi are competing with bacteria for nutrients and carbon sources. This competition has resulted in the ability to exploit various ecological niche by different microbes. Given a lower surface area to organism size, the fungi are able to deal with nutrient (N, P) scarcity better than any bacteria. They are very efficient in the use of nutrients to build new cell mass. Fungi could offer this benefit over bacteria in wastewater treatment processes. The biomass produced during fungal wastewater treatment has, potentially, a much higher value than that from the bacterial ASP. The fungi can be used to derive valuable biochemicals and can also be used as a protein source. Various high-value biochemicals are produced by commercial cultivation of fungi under aseptic conditions using expensive substrates. Several investigations have reported melanoidins degradation and distillery effluent decolorization

TABLE 5.3

Decolorization of Distillery Effluent and Reduction of Pollution Parameters by Pure Bacterial Strains as Reported by Different Group of Researchers

Bacterial Strain	Effluent	Abs.	BOD (%)	COD (%)	Incub	pH	Decol (%)	Temperature (°C)	Reference
Klebsiella oxytoca, Serratia mercescens, Citrobacter sp., and unknown bacterium	Melanoidins-containing wastewater	–	–	–	2 d	–	18.3	–	Jiranuntipon et al. (2008)
Pseudomonas aeruginosa PAO1, *Stenotrophomonas maltophila,* and *Proteus mirabilis*	Anaerobically treated distillery spent wash	–	–	–	72	–	67	37	Mohana et al. (2007)
Acinetobacter sp., *Pseudomonas* sp., *Comamonas* sp., *Klebsiella oxytoca, Serratia marcescens,* and unidentified bacteria, whereas anaerobically enriched consortium consisted of *Pseudomonas* sp., *Klebsiella oxytoca, Bacillus cereus,* and *Citrobacter farmeri,* a mercury-resistant bacterium and an unidentified bacterium	Molasses based distillery wastewater	–	–	–	48 h	–	26.5	–	Jiranuntipon et al. (2009)
Bacillus licheniformis, Bacillus sp., and *Alcaligenes* sp.	Synthetic and natural melanoidins	–	–	–	–	–	–	–	Bharagava et al. (2009)
Alcaligenes faecalis and *Bacillus cereus*	Sucrose glutamic acid-Maillard reaction product	–	–	–	–	–	–	–	Chandra et al. (2009)
Bacillus species, *Raoultella planticola,* and *Enterobacter sakazakii*	Sucrose aspartic acid- Maillard product	–	–	–	144 h	–	60–75	35	Yadav et al. (2011)
Proteus mirabilis, Bacillus sp., *Raoultella planticola,* and *Enterobacter sakazakii*	Biomethanated distillery spent wash	–	–	–	10 d	71	75	–	Yadav and Chandra (2012)
Klebsiella pneumoniae, Salmonella enterica, Enterobacter aerogenes, and *Enterobacter cloaceae*	Sucrose glutamic acid-Maillard reaction product	–	–	–	192 h	–	70	37	Kumar and Chandra (2018a)
Klebsiella pneumoniae, Salmonella enteric, Enterobacter aerogenes, and *Enterobacter cloacae*	Molasses-melanoidin	–	–	–	168	–	81	35	Chandra et al. (2018a)
Staphylococcus saprophyticus, Alcaligenaceae sp.	Anaerobically treated distillery effluent	–	–	–	–	–	–	–	Chowdhary et al. (2020)
Pseudomonas aeruginosa and *Bacillus cereus*	Distillery spent wash	–	–	66–78.66	21 d	–	–	–	Nayak et al. (2018)

SGA-MRP: sucrose glutamic acid-Maillard reaction products; MM: molasses-melanoidin; SAA-MP: sucrose-aspartic acid Maillard product; d: days; h: hours

TABLE 5.4

Decolorization of Distillery Effluent and Reduction of Pollution Parameters by Mixed Bacterial Strains as Reported by Different Group of Researchers

Bacterial Community	Effluent	Molecular Technique	Reference
Pseudomonas, Enterobacter, Stenotrophomonas, Aeromonas, Acinetobacter, and *Klebsiella*	Anaerobically treated molasses spent wash	ARDRA and BOX-PCR	Ghosh et al. (2004)
Stenotrophomonas, Enterobacter, Pantoea, Acinetobacter, and *Klebsiella* sp.	PMDE	PCR-RFLP	Chandra et al. (2012)
Microbacterium hydrocarbonoxydans, Achromobacter xylosoxidans, Bacillus subtilis, B. megaterium, and *B. anthracis* from upper zone; *B. licheniformis, A. xylosoxidans, Achromobacter* sp., *B. thuringiensis, B. licheniformis,* and *B. subtilis*	Distillery effluent contaminated site	16S rRNA	Chaturvedi et al. (2006)
Bacillus sp. and *Enterococcus* sp.	PMDS	PCR-RFLP	Chandra and Kumar (2017b)
Bacillus and *Stenotrophomonas*	DSW	PCR-RFLP	Chandra and Kumar (2017a)

ARDRA: amplified ribosomal DNA restriction analysis; PMDS: post-methanated distillery sludge; PMDE: post-methanated distillery effluent; DSW: distillery spent wash; PCR-RFLP: polymerase chain reaction-based restriction fragment length polymorphism; 16S rRNA: 16S ribosomal ribonucleic acid

by different species of fungi such as *Geotrichum candidum* (Kim and Shoda 1999), *Flavadon lavus* (Raghukumar and Rivonkar 2001), *Phanerochaete chrysosporium* (Thakkar et al. 2006), *Trametes* sp. (Gonzalez et al. 2000 2008), *Coriolus hirsutus* (Miyata et al. 2000), *Pleurotus florida, Aspergillus flavus* (Pant and Adholeya 2009a), *Neurospora intermedia* (Kaushik and Thakur 2013), *Fusarium verticillioides* (Pant and Adholeya 2009b), *Pleurotus sajor-caju* (Romanholo Ferreira et al. 2011; Aragao et al. 2020), *Trametes versicolor* (Rioja et al. 2008), and yeast *Citeromyces* sp., *Candida tropicalis* (Tiwari et al. 2012), *Candida glabrate* (Mahgoub et al. 2016), *P. chrysosporium* JAG-40 (Dahiya et al. 2001), *A. oryzae* MTCC 7691 (Chavan et al. 2013), and *Aspergillu fumigatus* UB$_2$60 (Ziaei-Rad et al. 2020). Fungal cultures belonging to white rot fungi such as *Phanerochaete* sp. have been extensively studied to develop bioprocesses for the mineralization of melanoidins. White-rot fungi are a class of microorganisms that produce efficient enzymes in extracellular, non-specific, and non-stereo selective enzyme system, including lignin peroxidase (LiP), laccase, and manganese peroxidase (MnP) capable of decomposing extremely diverse range of bio-refractory and toxic environmental pollutants, such as melanoidins, phenolics polyphenols, caramel, an alkaline degradation product of hexose (ADPH), and decolorized distillery effluent under aerobic conditions. Moreover, white-rot fungi also produce various oxidoreductases that degrade lignin and related aromatic compounds. White rot fungi do not require preconditioning to particular pollutants, because enzyme secretion depends on nutrient limitation rather than presence of pollutant. The extracellular enzyme system also enables white rot fungus to tolerate high concentration of pollutants. In relation to melanoidins degradation and effluent decolorization, white rot fungus *P. chrysosporium* is extensively studied (Kumar et al. 1998); however, some others too have gained significant attention, such as *Coriolus versicolor* (Kumar et al. 1998) and *Trametes* sp. (Gonzalez et al. 2000). *P. chrysosporium* emerges as an efficient remediation model for treating the effluent from textile, pulp and paper, and distilleries industries that contain refractory organic pollutants. Fungi typically grow in attached mode that is used for both support and as a substrate for growth and further produces environmentally important enzymes. Crude distillery effluent having a high COD value of 87,433 mg/L was used to produce bacterial cellulose under static fermentation by *Komagataeibacter saccharivorans* (Gayathri and Srinikethan 2019); 1.24 g/L of cellulose production was noted after 8 days along with 23.6% reduction in COD value. This study demonstrated that distillery effluent waters could be effectively reused as production medium fulfilling two objectives, namely, one reducing COD and making the effluent safe for disposal and two to produce a value-added product. Similarly, an attempt was made to utilize the distillery effluent for the production of bacterial cellulose by a novel bacterial species, *Gluconacetobacter oboediens* (Jahan et al. 2018). Maximum bacterial cellulose production of 0.85 g/100 mL was achieved in crude distillery effluent. The production was successfully scaled up to 1.0 L size producing 8.1 g of bacterial cellulose. Xylanase production by a newly isolated strain of *Burkholderia* sp. has been studied under solid-state fermentation using anaerobically treated distillery spent wash (Mohana et al. 2008). Hoarau et al. (2020) improved the production of single cell oil using the oleaginous yeast *Yarrowia lipolytica* MUCL30108 on distillery spent wash and to determine the main factor influencing biomass production. Taskin et al. (2016) studied the invertase production and molasses decolorization by cold-adapted filamentous fungus *Cladosporium herbarum* ER-25 in non-sterile molasses medium. The maximum invertase activity (36.1 U/mL) was attained after 72 h under the optimized non-sterile culture conditions. On the other hand, the fungus could remove toxical dark brown pigments (melanoidins) in non-sterilized molasses medium through biodegradation and bioadsorption mechanisms. A color removal rate of 64.8% in non-sterile medium could be achieved at the end of 144-h cultivation period. Authors found that Lac and MnP were responsible for biodegradation. Maximum laccase (4.6 U/mL) and MnP (3.5 U/mL) activities could be reached after 120 h. A single factorial experimental design was performed for decolorization of anaerobically treated distillery spent wash using *Cladosporium cladosporioides* (Ravikumar et al. 2011). Authors achieved a maximum color reduction of 62.5% and COD at optimum conditions. The optimum conditions required for the growth of the fungus was found to be 5 g/L of fructose, 3 g/L of peptone, 5 pH, and 35 °C. It was also observed that during the process, a maximum of 1.2 g of fungal growth was attained. Biodegradation and decolorization of distillery spent wash by a novel strain *C. cladosporioides* has been reported (Ravikumar et al. 2013). Scientists conducted a 24 full factorial central composite experimental design of Response Surface Methodology to obtain a maximum decolorization of 62.5% and COD reduction of 73.6% at optimized conditions (fructose concentration of 7 g/L, peptone 2 g/L, 6 pH, and 10% (w/v) inoculum concentration). Ahmed et al. (2018) reported the sustainable bioremediation of sugarcane vinasse using autochthonous basidiomycetous fungi *Pycnoporus* sp. P6 and *Trametes* sp. T3. In liquid cultures, both fungi demonstrated their capacity to decolorize and remove phenolic compounds from diluted vinasse, with the concomitant production of ligninolytic enzymes, mainly laccases, suggesting the participation of this enzyme in the bioremediation process. Ziaei-Rad et al. (2020) studied biodecolorization mechanism of effluents from ethanol production plants using *Aspergillus fumigatus* UB260. A color removal rate of 47% was obtained at an incubation time of 96 h at 30 °C. SEM micrographs showed the accumulation of specified pigments in the mycelia. The release of color from fungal mycelia to 0.5 N NaOH proved the color adsorption on the surface of mycelia. Poly-acrylamide gel electrophoresis analysis (SDS/PAGE) proved the presence of protein molecules in bio-decolorization culture. Treatment of sugarcane vinasse using an autochthonous fungus *Aspergillus* sp. V2 from the northwest of Argentina and its potential application in fertigation practices has been evaluated by Rulli et al. (2020). In order to evaluate the application of vinasse in fertigation, phytotoxicity of the residue, before and after biological treatment, was tested on two plant species used for human consumption (*Triticum aestivum* L. and *Raphanus sativus*). The results of this study showed that *Aspergillus* sp. V2 is useful for the recovery of an actual wastewater like sugarcane vinasse, enhancing its fertilizing properties. Numerous successful studies showing degradation and decolorization of distillery effluent by fungal treatment are summarized in **Table 5.5**.

The utilization of distillery wastewater as a feasible culture medium for fungi is hindered by the presence of inhibitory compounds, such as phenols and other recalcitrant structures. The effect of phenolic acids and molasses spent wash concentration on distillery wastewater remediation by four fungi (*G. candidum, C. versicolor, P. chrysosporium*, and *Mycelia sterilia*) has been evaluated by FitzGibbon et al. (1998). Fungal growth was inhibited to a varying extent in the presence of gallic and vanillic acid, except for *G. candidum*, which was unaffected by gallic acid. *G. candidum* and *P. chrysosporium* growth rates increased in the presence of increasing concentrations of molasses spent wash (up to 50% v/v); however, growth of *M. sterilia* and *C. versicolor* was inhibited at spent wash concentrations above 5% (v/v). Increasing the concentration of MSW from 6.25% (v/v) to 12.5% (v/v) increased the decolorizing ability of each

TABLE 5.5

List of Identified Bacterial Communities Grown in Distillery Waste Contaminated Environment by Various Molecular Techniques

Type of Wastewater	Species	Abs.	Reduction of Various Pollution Parameters (% Max.)						Mel. Rem.	Reference
			BOD (%)	COD (%)	Decol (%)	Incub	pH	Temperature (°C)		
Melanoidin-containing wastewater	*Coriolus hirsutus*	—	—	—	—	—	—	—	—	Miyata et al. (2000, 1998)
Distillery effluent	*Fusarium verticillioides* TERIDB16, *Alternaria gaisen* TERIDB6	475	—	—	37	—	—	—	—	Pant and Adholeya (2009)
ATDSW	*Cladosporium cladosporioides*	—	—	62.5	52.6	—	5	35	—	Ravikumar et al. (2011)
ATDSW	*Coriolus versicolor* and *Phanerochaete chrysospotium*	—	—	90.0 and 73.0	71.5 and 53.5	10 d	5	35–40	—	Kumar et al. (1998)
ATDSW	*Aspergillus* species	—	—	—	—	—	6	—	—	Wagha and Nemade (2018)
Distillery effluent	*Aspergillus niveus*	—	94	97.14	56	—	9.0	—	—	Angayarkanni et al. (2003)
Distillery effluent	*Pleurotus florida* EM 1303 *P. pinophilum* TERI DB1 and *A. gaisen* TERI DB6	475	—	—	86, 50, and 47	—	—	—	—	Pant and Adholeya (2007)
Distillery mill effluent	*Emericella nidulans* var. lata and *Neurospora intermedia*	456	—	—	38–62 (DF3); 31–64 (DF4)	—	3	30	—	(Kaushik and Thakur, 2009)
Distillery spent wash	*Aspergillus oryzae* MTCC 7691	—	51	86.19	75.71	25 d	—	30	—	Chavan et al. (2013)
Distillery wastewater	*Pycnoporus coccineus* strains	—	—	—	—	—	—	43	—	Chairattanamanokorn et al. (2005)
Distillery wastewater	*Aspergillus fumigatus* UB260	—	—	—	—	—	—	—	—	Pazouki et al. (2017)
Distillery wastewater	*Trametes* sp. 1-62	—	—	61.7	73.3	7 d	—	—	—	Gonzalez et al. (2000)
Melanoidin solution	*Coriolus* sp. No. 20	450	—	—	—	2–5 d	—	—	80	Watanabe et al. (1982)
Melanoidin solutions	*Candida glabrata*	400–800	—	70	60	24 h	5	30	—	Mahgoub et al. (2016)
Melanoidin-containing wastewaters	*Trametes* sp. 1-62	—	—	—	—	—	4	—	—	Gonzalez et al. (2008)

(Continued)

TABLE 5.5 *(Continued)*

Type of Wastewater	Species	Reduction of Various Pollution Parameters (% Max.)								Reference
		Abs.	BOD (%)	COD (%)	Decol (%)	Incub	pH	Temperature (°C)	Mel. Rem.	
Molasses distillery wastewater	*Pleurotus ostreatus (Florida) Eger EM 1303, Aspergillus flavus TERI DB9, Fusarium verticillioides ITCC 6140 Aspergillus niger TERI DB18, and Aspergillus niger TERI DB20*	475	–	86.33, 74.67, 68.33, 64.67, and 54.67	–	28 d	–	–	–	Pant and Adholeya (2007)
Molasses spent wash	*Flavodon flavus (Klotzsch) Ryvarden*	–	–	–	80	8 d	–	–	–	Raghukumar and Rivonkar (2001)
Molasses spent wash	*Phanerochaete chrysosporium ATCC 24725*	–	–	–	85	10 d	–	–	–	Fahy et al. (1997)
Molasses wastewater	*Coriolus versicolor, Funalia trogii, Phanerochaete chrysosporium, and Pleurotus pulmonarius (Pleurotus sajor-caju)*	475	–	–	–	–	–	–	–	Kahraman and Yeilada (2003)
Molasses wastewater	*Penicillium decumbens*	–	–	50.7	40	1-5 d	5.2	–	–	Jimenez et al. (2003)
Molasses wastewater	*Issatchenkia orientalis No. SF9-246*	475	77.4	80.0	75–80	7 d	5	30	–	Tondee et al. (2008)
Raw vinasse	*Pleurotus sajor-caju*	475	–	–	92	–	–	–	–	Junior et al. (2020)
Spent wash	*Cladosporium cladosporioides*	–	–	73.6	62.5	5.02 d	6	–	–	Ravikumar et al. (2013)
Spent wash	*Candida tropicalis RG-9*	475	–	–	75	24 h	–	45	–	Tiwari et al. (2012)
Stillage	*Citeromyces sp. WR-43-6*	475	76	100	75	8 d	–	60–70	–	Sirianuntapiboon et al. (2004)
Sugarcane distillery wastewater	*Aspergillus niger*	–	–	–	–	–	–	–	–	Chuppa-Tostain et al. (2018)
Synthetic and spent wash melanoidins	*Phanerochaete chrysosporium JAG-40*	–	–	–	–	6 d	–	30	80	Dahiya et al. (2001)
Vinasse	*Pleurotus sajor-caju CCB 020*	–	–	–	–	–	–	–	–	Ferreira et al. (2011)
Vinasse	*Trametes hirsuta Strain Bm-2*	–	–	–	69.2	–	–	–	–	Tapia-Tussell et al. (2015)
Vinasse	*Trametes versicolor*	475	–	38	–	–	–	–	–	España-Gamboa et al. (2013)

Waste/substrate	Microorganism	Abs (475)	BOD removal	COD removal	Decol	Incub	pH	Temp	Mel Rem	Reference
Molasses wastewater	*Aspergillus fumigatus* G-2-6	475	–	–	75	3–4 d	–	45	–	Ohmomo et al. (1987)
Melanoidin-containing solution	*Aspergillus oryzae* Y-2-32	475	–	–	–	4 d	–	35	75	Ohmomo et al. (1988)
Molasses melanoidin and synthetic melanoidins media	*Rhizoctonia* sp. D-90	475	–	–	–	8 d	–	–	87.5 and 84.5	Sirianuntapiboon et al. (1991)
Alcohol distillery wastewater	*Pycnoporus coccineus*	–	–	–	–	–	–	35 and 43	–	Chairattanamanokorn et al. (2005)
Molasses spent wash	*Phanerochaete chrysosporium*	–	–	–	85	10 d	–	–	–	Fahy et al. (1997)
Anaerobically treated distillery spent wash	*Cladosporium cladosporioides*	–	–	62.5	52.6	–	–	35	–	Ravikumar et al. (2011)
Melanoidins	*Candida tropicalis* RG-9	–	–	–	75	24 h	–	45	–	Tiwari et al. (2012)
Distillery wastewater	*Trametes sp.* I-62	–	–	61.7	73.3	7 d	–	–	–	González et al. (2000)
Model melanoidin (glucose and glycine)	*Streptomyces werraensis* TT 14	–	–	–	64	–	–	–	–	Murata et al. (1992)
Distillery wastewater	*Aspergillus fumigatus* UB260	–	–	–	–	–	–	–	–	Pazouki et al. (2008)
Distillery effluent	*Aspergillus niveus*	–	94	97.14	56	–	–	–	–	Angayarkanni et al. (2003)
Anaerobically digested spent wash	*Coriolus versicolor; Phanerochaete chrysospotium*	–	90 and 73	–	71.5 and 53.5	–	–	35–40	–	(Kumar et al., 1998)
Aspergillus species	Anaerobically treated distillery spent wash	–	–	–	–	–	–	–	–	Wagh and Nemade (2018)
Geotrichum candidum	Spent wash	–	–	53.17	–	10 d	–	30	–	FitzGibbon et al. (1995)
Coriolus versicolor	Spent wash	–	–	77	4.2	10 d	–	30	–	Fitzgibbon et al. (1995)

ATDSW: anaerobically treated distillery spent wash; Abs: Absorbance; BOD: biological oxygen demand; COD: chemical oxygen demand; Decol: decolorization, Incub: incubation; d: days; h: hours; Mel. Rem: melanoidin removal.

fungus, except for *M. sterilia. C. versicolor* exhibited greatest color removal with a reduction of 0.43 units at A_{475} (equivalent to 53% color reduction) after 10 days growth in 12.5% (v/v) MSW. Rodrigues Reis et al. (2020) analyzed the effect of different homogeneous AOPs to lower the concentration of inhibitory phenolic compounds in vinasse, and to test the effects of the pretreatment on the growth of an oleaginous fungal strain. Vinasse was treated using a batch system with the addition of ferrous ion (Fe^{2+}) (Fenton), in the presence of UV, and with the addition of H_2O_2. All of the reaction systems were able to partially degrade the COD, ranging from 18.4% up to 54.1% in the factorial results. The pretreated vinasse samples were assayed for the growth performance of *Mucor circinelloides*, demonstrating that the control value of 3.54 g/L was increased to values as high as 4.84 g/L, also followed by an increase of ≈73% in terms of lipid productivity. Kaushik et al. (2010) successfully treated distillery spent wash by *Emericella nidulans* var. lata, *N. intermedia* and *Bacillus* sp. in a three-stage bioreactor. Aragão et al. (2019) evaluated the degradation of sugarcane vinasse with the production of biomass by *P. sajor-caju* CCB020, considering the combination of temperature and pH effects, using surface response methodology (RSM). The optimum temperature and pH values were respectively 27 °C and 5.6 for maximum decolorization yield and 20 °C and 6.8 for maximum biomass production. In parallel, scale-up experiments under conditions of 30 °C and initial pH 5 were evaluated in two different air-lift bioreactors of 7.0 L. Under these conditions, reductions of 53% and 58% in COD and 71% and 58% in BOD were obtained, respectively, with the concentric tube-type air-lift bioreactor with an increased degassing zone and without an increased degassing zone. When compared to the treatments evaluated with individually grown microorganisms, the mixed treatment proposed produced the best COD results in effluent degradation. Velásquez-Riaño et al. (2018) proposed an innovative biological alternative in degradation processes of great organic pollutants such as vinasse. They evaluated the capacity of a mixed treatment with *Komagataeibacter kakiaceti* GM5 and *T. versicolor* DSM 3086 to degrade and reduce the toxicity of this by-product. However, better results were obtained when using the treatment with *T. versicolor* DSM 3086 for color removal and treatment with *K. kakiaceti* GM5 produced better results in terms of turbidity and toxicity. Although the simultaneous mixed treatment did not improve all the variables studied, it never showed the lowest performance. Many of the decolorization studies with fungi had been conducted using fungal mycelia in the live or dead form. However, the direct application of fungus for the bioremediation has some disadvantages. One of the major disadvantages of using fungal cultures is the accumulation of biomass, which would cost the wastewater treatment on industrial scale. However, the performance of fungal decolorization was limited by long growth cycle and moderate decolorization rate. Most fungi do not do well in suspended growth as found in the water column. This is where bacteria are very efficient at uptake of nutrients, carbon, and oxygen. However, fungal treatments have not yet found a real application, mainly due to the difficulty in selecting organisms able to grow and remain active in the very variable and harsh conditions of wastewaters. In fact, fungi should be able to live with the scarce nutrient resources present in the effluents, to win the competition with the autochthonous microflora, and to survive in the presence of high concentrations of salts, dyes, detergents, and heavy metals. Moreover, the requirement of low pH values (4–5), optimum enzyme activity, and long hydraulic retention times for complete decolorization are other concerns in fungal treatments application.

5.4.1.5 Yeast-Assisted Treatment

Yeast has mainly been studied with regard to biosorption. Compared to bacteria and filamentous fungi, yeast has many advantages; they not only grow rapidly like bacteria, but like filamentous fungi, they also have ability to resist unfavorable environments. To ascertain the decolorizing potential of yeasts, minor work has been done, especially focusing on the biodegradation. Few yeast species, for instance, *C. tropicalis* (Tiwari et al. 2012), *Issatchenkia orientalis*, and *C. glabrata* (Mahgoub et al. 2016), are utilized for assessing the enzymatic biodegradation accompanied with the decolorization of melanoidins. A strain identified as *Issatchenkia orientalis* showed the highest potential for decolorization of anaerobically treated molasses wastewater with melanoidin pigment; decolorization, COD and BOD removal rates reached up to 91%, 80%, and 77%, respectively, at 30 °C and pH 5 during a 7-day batch culture time (Tondee and Sirianutapiboon 2006; Tondee et al. 2008). The thermotolerant *C. tropicalis* decolorized complex melanoidin compounds at a wide range of temperature and pH conditions, in the presence of small amounts of carbon and nitrogen sources, within a short incubation period of 24 h. This strain could reduce environmental pollution problems due to molasses wastewater discharge by decolorizing melanoidin pigment under a cost-effective and eco-friendly process (Tiwari et al. 2012). More recently, a study has shown that yeast species *Saccharomyces cerevisiae* (CCMA0187 and CCMA0188) and one strain of *Candida glabrata* (CCMA0193) and *Candida parapsilosis* (CCMA0544) acted as a promising agent in reduction of vinasse toxicity or a decrease of 55.8% and 46.9% in BOD and COD, respectively.

5.4.2 Biocomposting

Composting is a biological disintegration process in which organic wastes/matters are transformed into humus-like matters by mixed population of microbes like bacteria, fungi, actinomycetes, and protozoa, which is a stable organic end product called as compost. In composting, the contaminated soil is dug out and blended with a bulking agent and organic materials (OMs; i.e., animal wastes, industrial sludge, wood chips, vegetative wastes, etc.) (Figure 5.5). The existence of these organic constituents aids the proliferation of a rich microbial community that changed the organic matter into compost via their enzymatic activity. Usually composting is an anaerobic, thermophilic procedure of microbiological disintegration of polluting agents (organic wastes) into compost, which can be disposed safely into the environment. Under normal environmental circumstances, earthworms, soil insects, and nematodes start the degradation of OMs into minute particles, thus intensifying their bioavailability for the microbial communities, whereas under regulated environmental conditions, composting machinists disintegrate the large waste entities via chopping or grinding. During composting, microbes degrade the organic compounds to acquire energy for carrying out the metabolic activities and obtain nutrients for their survival and growth. The molasses-based biomethanated distillery wastewater is presently effectively utilized with sugarcane press mud through the composting process. The composting process involves an interaction between the organic waste and microorganisms. Biocomposting of spent wash with sugarcane press mud is a common practice in sugar industries, where press mud or filter cake is another commonly known major waste of the sugar industry, a spongy amorphous and dark brown to brownish white material, which contains sugar, fiber, lignin, cellulose, inorganic salts, and soil particles (). These organic wastes when applied without composting to the soil may lead to temporary lock up of nutrients as a result of impaired C:N ratio and may not be beneficial to crop. Hence, composting of organic wastes is necessary to reduce the lignin and cellulose contents, thereby the nutrient availability is improved. In practice, the composting process is slow due to the low biodegradability and recalcitrant nature of biomethanated distillery spent wash and necessitates high retention time. The biocomposting system helps distilleries to utilize the sludge materials segregated from the distillery processes, incineration, spent wash, and RO. Plant rejects to convert it into valuable biocompost using sugar unit press mud. The falling population of livestock has brought down the availability of Farm Yard manure. Biocompost is now filling up this gap as an organomineral supplement. Biocompost prepared from distillery spent wash was reported to contain higher organic carbon (15.5%), N (2%), P (2.5%), and K (3%). The pH of the compost was found ideal (7–7.5) with a C:N ratio of 15:1. Application of biocompost and 50% NPK application was found to have enhanced the available N, P, and K status of soil and recorded maximum yields of cane over the yield when 100% NPK was used (Rakkiyappan et al. 2001). Spent wash-press mud compost has been observed to improve the stability of aggregates and porosity. Composting also assists in the degradation of colored organics in the distillery effluents, which also enrich the compost with nutrients, especially potassium. In order to provide a balanced nutritional value and enrich it more, the compost could be enriched with the use of rock phosphate, gypsum, yeast sludge, bagasse, sugarcane trash, boiler ash, coir pith, and water hyacinth. Both aerobic and anaerobic composting techniques have been suggested requiring about 30 days for active reactions and another 30 days for maturing. Spent wash utilization in aerobic composting is more than spent wash consumptions in anaerobic composting. The application of composted sugar industry wastes in the form of biocompost not only enhances the soil nutrient status and cane yield but also serves as a means for eco-friendly management of industrial by-products. The above fertilizer is produced from organic matter substituting the imported chemical fertilizers that are highly subsidized. Since distilleries are going for concentration of SW used for compost preparation, there is a need to reevaluate the utilization rate of the SW and quality of the compost.

FIGURE 5.5   A site view of composting of distillery waste.

5.4.3 Phytoremediation of Distillery Waste

Phytoremediation, also referred to as botanical bioremediation or green remediation, is an environment-friendly approach, which is mainly concerned with the usage of green plants and their associated microbial communities, to improve the quality of environment and prevention of pollution discharge in the environment. Phytoremediation process is simpler, cost-effective, green, and environmentally safe when compared to other biological methods (Chandra et al. 2015; Chandra and Kumar 2017c; Kumar et al. 2018; Chandra et al. 2018c). Plants are ecologically safe, economically beneficial, and robust renewable resource for in situ reduction of organic and inorganic contaminants present in distillery waste. The introduction of plants to the polluted site(s) has the potential to yield several indirect contaminant attenuation mechanisms, which assist in the removal of toxic substances/management of polluted sites (Chowdhary et al. 2018b; Chandra et al. 2018b). Depending on the type of contaminants, the site conditions, and the level of cleanup required, phytoremediation technology is currently divided into a number of processes, namely, phytoextraction, phytofiltration, phytostabilization, phytovolatilization, and phytodegradation, with each process having a different mechanism of action for remediation of environmental pollutants from the contaminated site (Chandra et al. 2018c; Kumar and Chandra 2020a). Phytoextraction refers to the use of pollutants-accumulating plants to absorb, translocate, and store metals or organic contaminants from soil, sediments, sludge, water, and/or wastewater in their root and shoot tissues. The term phytoextraction is mostly used to refer to removal of metal from contaminated matrices (Chandra and Kumar 2018b). Phytodegradation, also referred to as phytotransformation, is the breakdown of contaminants taken up by plants through metabolic processes within the plant, or the breakdown of contaminants surrounding the plant through the effect of enzymes produced by the plants (Kumar et al. 2021c). Plants are able to produce enzymes that catalyze and accelerate degradation. Hence, organic pollutants are broken down into simpler molecular forms that are integrated with plant tissue, which in turn foster plant growth. The release of volatile contaminants to the atmosphere via plant transpiration, called phytovolatilization, is a form of phytotransformation (Kumar et al. 2021b). Although transfer of contaminants to the atmosphere may not achieve the goal of complete remediation, phytovolatilization may be desirable in that prolonged soil exposure and the risk of groundwater contamination is reduced. Rhizofiltration, a root-zone in situ or ex situ technology, can be used for the elimination of metals from water and aqueous waste streams that are retained only within the roots of aquatic plants. It reduces the mobility of metals and prevents their migration to the groundwater, thus reducing bioavailability for entry into the food chain. Rhizofiltration is particularly effective when low concentrations of contaminants and a large volume of water are involved. However, plants and their partner microorganisms are often hampered by unavailability of contaminants to the rhizospheric microbes due to rapid uptake into the plants as well as phytotoxicity to plants due to increased accumulation in plant tissues. A promising alternative consists of optimizing the synergistic effect of plants and microorganisms by coupling phytoextraction with soil bioaugmentation, also called rhizoremediation. This technique has been widely developed for the remediation of soils contaminated by organic pollutants but not for metals. Rhizospheric bacteria are crucial for the success of phytoremediation. Rhizospheric bacteria can improve phytoremediation through altering metal bioavailability by changing the environmental pH, producing chelating substances, and altering redox potentials (Chandra and Kumar 2015b; Kumar and Chandra 2018c; 2020c). Bacterium may utilize different mechanisms under different environmental conditions to promote plant growth at polluted sites. Successful phytoremediation depends not only on the interaction of plant roots with bacteria, but also on the bioavailability of heavy metals in soils. Chandra et al. (2018b) stated that heavy metal phytoextraction via native weeds and grasses is an emerging practice ensuring the effective and cheap way to remediate the contaminated sludge or soil that were polluted with organic contaminants and heavy metals. Though plants play a major role in the distillery waste remediation, there are problems that limit their use in this scenario such as the level of contaminants tolerated by the plant, the bioavailable portion of pollutants, evapotranspiration of unstable organic contaminants, and necessitating great zones for grafting the approach.

5.5 Combined/Hybrid Treatment

Biological methods (mainly based on conventional activated sludge systems) are widely applied, but they are not adequate enough for the efficient removal of color and recalcitrant pollutants present in distillery effluents, neither as on-site nor as a centralized treatment. Thus, integration/combination of different treatment methodologies are more promising in reducing the physicochemical parameters, metal ions, as well as organic compounds from distillery effluent. A study on the synergy of a combined anaerobic digestion and UV photodegradation treatment of distillery effluent was carried out in fluidized bed reactors to evaluate pollution reduction and energy utilization efficiencies by Apollo and Aoyi (2016). The combined process improved color removal from 41% to 85% compared to that of anaerobic digestion employed as a stand-alone process. An overall corresponding TOC reduction of 83% was achieved. The bioenergy production by the anaerobic digestion step was 14.2 kJ/g TOC biodegraded, while UV lamp energy consumption was 0.9 kJ/mg TOC, corresponding to up to 100% color removal. Electrical energy per order analysis for the photodegradation process showed that the bioenergy produced was 20% of that required by the UV lamp to photodegrade 1 m^3 of undiluted pre-AD-treated effluent up to 75% color reduction. It was concluded that a combined AD-UV system for treatment of distillery effluent is effective in organic load removal and can be operated at a reduced cost. A hybrid nanofiltration (NF) and RO pilot plant has been reported to remove the color and contaminants of the distillery spent wash. Color removal by NF and a high rejection of 99.80% TDS, 99.90% of COD, and 99.99% of potassium was achieved from the RO runs, by retaining a significant flux as compared to pure water flux, which shows that membranes were not affected by fouling during wastewater run. The ability to recover valuable materials, recycle water, reducing freshwater consumption and wastewater treatment costs, using a point-source approach for minimum capital costs, small disposal volumes that will minimize the waste disposal costs, and reduction in regulatory pressures and fines improved the heat recovery systems. The pyrolysis/gas chromatography/mass spectrometry results showed a decrease in a number of pyrolysis products after 7 days of fungal treatment, mainly furan derivatives. The decrease in the relative areas of these compounds could be related to the vinasse color removal associated with melanoidin degradation. Pyrolysis/gas chromatography/mass spectrometry was applied to the chemical characterization of several fractions of Cuban distillery wastewater as well as to monitoring the changes that occurred after fungal treatment with this white-rot basidiomycete. Maximum effluent decolorization values and COD reduction attained after 7 days of fungal treatment were 73.3% and 61.7%, respectively, when 20% (v/v) of distillery vinasses was added to the culture medium. Under these conditions, a 35-fold increase in laccase production by *Trametes* sp. I-62 was measured, but no manganese peroxidase activity could be detected (González et al. 2000). Sangave and Pandit (2004) sonicated distillery wastewater as a pretreatment step to convert complex molecules into a more utilizable form by cavitation. Samples exposed to 2 h ultrasound pretreatment displayed 44% COD removal after 72 h of conventional aerobic oxidation compared to 25% COD reduction shown by untreated samples. These results are contrary to those of Mandal et al. (2003), who concluded that ultrasonic treatment was ineffective for distillery spent wash treatment. The combined processes of sonolysis, enzymatic hydrolysis, and aerobic biological oxidation were found to increase the biodegradation efficiency of distillery wastewater (Sangave et al. 2007a,b). However, the timescale and the dissipated power necessary to obtain complete mineralization of the pollutants in the case of sonolysis are not economically acceptable. Hence, ultrasound is found more effective when used in combination with other conventional treatment processes than as a stand-alone process. Besides, the application of sonolysis in the treatment of complex effluents is limited by the high installation/ process costs as well as low COD removal. A hybrid pretreatment technique was explored for the selective improvement in biodegradability of biomethanated spent wash (Bhoite and Vaidya, 2018a,b). In this work, the biodegradability of biomethanated distillery spent wash (BOD 58 and 100 and COD 40,000 mg/L) was improved by oxidation using an FeSO$_4$ catalyst at 175 °C for 1 h. After oxidation, adsorption over activated carbon (loading 5 g/100 mL wastewater) resulted in 73% reduction in COD and a substantial increase in the BOD5/COD ratio (0.2–0.45). This BI further rose to 0.52 when anaerobic digestion was performed, using 1% acclimatized biomass. After a final polishing aerobic treatment step, the BI and COD were reduced up to 0.58% and 91%, respectively. Clearly, this work has provided a useful solution to the effective pretreatment and valorization of biomethanated distillery wastewaters. Additionally, this hybrid pretreatment technique provides the opportunity to generate more methane from the distillery spent wash (Bhoite and Vaidya 2018b). Pretreatment of molasses spent wash (vinasse) with vortex-based cavitation significantly enhanced biogas generation (Nagarajan and Ranade 2020). Kumar et al. (2006) carried out an experiment to explore strategies for the mineralization of refractory compounds in distillery spent wash by anaerobic biodegradation/ozonation/aerobic biodegradation. The treatment of distillery spent wash by anaerobic-aerobic biodegradation was shown in an overall COD removal of 70.8%. Ozonation of the anaerobically digested distillery spent wash was carried out as is (phase I experiments) and after pH reduction and the removal of inorganic carbon (phase II experiments). The ozonation step resulted in an increase in overall COD removal (95%) obtained when an ozone dose of approximately 5.3 mg ozone absorbed/mg initial total organic carbon was used. The author concluded that the combination of chemical and biological processes led to the greater destruction of organic contaminants present in the effluent (Kumar et al. 2006). Similarly, laboratory-scale experiments were performed to degrade highly concentrated organic matter in the form of color in distillery spent wash through batch oxidative methods such as electrocoagulation (EC), electro-Fenton (EF), and the Fenton process (David et al. 2015a,c).

5.6 Enzymes Involved in the Microbial Decolorization and Degradation of Melanoidins

Microbes are known to play a major role in degradation and detoxification of Maillard reaction products of the distillery effluent. These fully or some time partly decolorize distillery effluent due to their lignin-modifying enzymes, also termed as ligninolytic enzymes, such as LiP, MnP, and laccase. Many of the enzymes used in degrading the most recalcitrant portion of plant waste (lignin) also make these enzymes able to initiate decomposition of complex organics. Numerous bacteria, fungi, and yeast species could not only decolorize several azo dyes but also completely degrade them under certain ecological circumstances.

5.7 Utilization in Agriculture: The Most Sustainable Option

Distillery effluents are a rich source of nitrogen, phosphorous, and potassium. In addition, they also contain large amounts of Ca, Mg, Na, S, and chlorides that can be used as a resource for crop production and reduce the use of inorganic fertilizers (Chauhan and Rai 2010; Kumari et al. 2016). It also has appreciable quantities of micronutrients. Being organic in nature, the nutrients are more rapidly taken up by plants from soil. Distillery effluents have been found to be more effective than a mixture of inorganic fertilizers and cow dung manure. Due to its high organic content, effluent may serve as a good fertilizer for crops, more effective than inorganic or mixed fertilizers being used by farmers (Arora et al. 1992). The energy, fertilization, and irrigation potential of distillery effluents have helped the industry to build immense social acceptability now. Distillery effluents are used as a supplement to mineral fertilizer in Brazil. Nutrient recycling through the application of vinasse and filter cake to sugarcane crops in Brazil has reduced the consumption of fertilizers as compared to other crops and in other countries. In Australia, spent wash is blended with additional crop nutrients and sold as manure. Spent wash could also be used for composting the trash in fields. Ferti-irrigation are the most widely used options for effluent disposal. The utilization of vinasse in ferti-irrigation practices started in the 1950s, and by the 1980s it was a common practice for sugarcane refineries to utilize the liquid residual as fertilizer. The concept behind ferti-irrigation consists of a sum of irrigation to sugarcane fields, by the percolation of vinasse liquid to the soil, with the simultaneous fertilization, transferring its nutrients to the plants. Some authors have observed that spent wash irrigation may also lower the incidence of insect pests. Besides decreasing the costs involved with chemical fertilizers, completely supplying phosphorus, and being of low capital cost, vinasse utilization in ferti-irrigation practices could be considered of certain level of environmental concern. Ferti-irrigation practices have been linked with increase in eutrophication of water bodies and the formation of dead aquatic bodies in Brazil and in other countries. The high toxicity potential of vinasse being utilized in ferti-irrigation practices may lead to hydrologic, agronomic, and social problems. Since there are no pollution charges applied to sugarcane farmers and ethanol producers, ferti-irrigation still stands as the predominantly application of vinasse.

6

Ligninolytic Enzymes in Degradation and Detoxification of Distillery Waste

6.1 Introduction

Ethanol-producing distillery industry is one of the oldest industries in India. It is typically characterized as pollutants-generating industries that produce large quantities of colored effluents during the ethanol production. Distillery effluent is recognized as a serious environmental threat due to high chemical levels, including pH, organic load (chemical oxygen load or demand, and biological oxygen demand), and inorganic load. Melanoidins, a major colorant of distillery effluent, are non-degradable and resistant to light and highly toxic to aquatic life and soil microorganisms. Therefore, environmental regulations in most countries require effluent to be decolorized before its discharge. The complex molecular structure of the melanoidins in the effluents makes it difficult to be removed by conventional wastewater treatment process because physical and chemical treatment process generates huge amounts of sludge, which leads to secondary pollution and formation of hazardous by-products and secondary elements and is also inefficient due to capital cost and operating speed. In recent years, there has been considerable interest in finding innovative and environment-friendly treatment technologies to substitute the conventional process. Many studies have shown the degradation and detoxification of distillery effluent by microbial culture, which, when cultured under appropriate conditions, produce extracellular ligninolytic enzymes, including laccase (Lac), manganese peroxidase (MnP), and lignin peroxidases (LiPs), which are capable of breaking a lot of different chemical bonds present in organic and inorganic substrates (Kumar and Chandra 2020d). A new group of peroxidases known as dye-decolorizing peroxidases (DyPs), which is also capable of degrading lignin, has been identified in fungi. Moreover, ligninolytic enzymes have a great potential in biotechnological applications, such as in the food industry, delignification of cellulosic compounds, paper bleaching, degradation of synthetic dyes and pesticides in the soil, and in the breakdown against several micropollutants, including already recognized endocrine-disrupting chemicals (EDCs) at their natural residual concentrations. White-rot fungi (WRF) and their ligninolytic enzymes are considered as a promising biotechnological tools to remove lignin, lignin monomers, melanoidins, Maillard reaction products, and other related organic pollutants from industrial effluent and contaminated ecosystems (Kumar and Chandra 2020d). Despite the use of WRF, ligninolytic bacteria are also promising candidates for degradation of melanoidins, plant phenolics, androgenic-mutagenic compounds, and EDCs, perhaps, because of their dexterity in the degradation of recalcitrant compounds and their abilities to produce some lignin-modifying enzymes, including Lac, MnP, and LiP. This chapter deals with the structure, composition, and mode of action of ligninolytic enzymes in degradation, decolorization, and detoxification of distillery effluent. Moreover, the role of various miscellaneous (non-ligninolytic) enzymes in biodegradation and decolorization of distillery effluent is discussed.

6.2 Ligninolytic Enzymes

The term ligninolytic enzymes, also known as lignin-degrading enzymes, encompasses four oxidative enzymes: Lac, MnPs, LiPs, and VPLs. LiP, MnP, and VP are heme peroxidases, which require H_2O_2 as

DOI: 10.1201/9781003029885-6

an oxidant for their catalytic activity. The demand for these enzymes has increased in the recent years due to their potential applications in diverse biotechnological applications. Ligninolytic enzymes are being widely used in the pollution abatement, especially in the treatment of industrial effluents that contain hazardous compounds like melanoidins, dyes, EDCs, phenolics, fatty acids, androgenic-mutagenic compounds, and other xenobiotics. Besides the lignin degradation activity of ligninolytic enzymes, they have shown great potentials for the distillery effluent decolorization and degradation of melanoidins and non-phenolic compounds has been extensively studied (Pant, and Adholeya, 2007a, b, 2009a; Kumar and Chandra 2018a; Chandra et al. 2018a). The resourcefulness of ligninolytic enzymes in the elimination of recalcitrant pollutants is, perhaps, due to their high redox potentials for the oxidation of a wide spectrum of organic compounds.

6.2.1 Laccases and Their Role in Distillery Effluent Treatment

Lac is blue monomeric or multimeric copper-containing polyphenol oxidases (E.C. 1.10.3.2; benzene-diol: oxygen oxidoreductase) that are widely distributed among microorganisms (fungi and bacteria), insects, and higher plants, showing a specific function in each of them. In nineteenth-century, Lac was first isolated from exudates of the Japanese tree *Rhus vernicifera*. Later, Lac have been mostly isolated and characterized from plants and fungi, but only fungal laccases are used currently in biotechnological applications for the degradation and detoxification of complex effluent. Structurally, Lac are monomeric, dimeric, and tetrameric glycoproteins, having fewer saccharide compounds in fungus and bacteria than in plant enzymes (Chandra et al. 2017). Mannose is one of the major components of the carbohydrates attached to Lac. The carbohydrates, which are 10–45% of the total molecular mass, are covalently linked, and due to this property of enzymes show high stability under stress environmental conditions. The molecular weight of a Lac is determined to be in the range of 50–97 kDa from various experimental reports (Kallio et al. 2010). The redox potential of microbial Lac ranges from 0.4 to 0.5 V, but they are active and stable at high temperatures (66 h at 60 °C), at pH 7–9, and high salt concentrations. Although some substances may inhibit Lac activity, the addition of certain compounds can enhance its efficiency. Likewise, it is found in the literature that the Lac are characterized by the presence of copper (Cu) centers inside their catalytic core. In the catalytic center of each of them, there are four Cu atoms distributed in three different sites, with T1, at least three additional copper ions: T2 and T3 copper ions, arranged in a trinuclear cluster. The reactive surface exposed Cu center (type I copper T1) formed by one ion and is responsible for substrate oxidation and electron transfer and gives the characteristic blue color; and two buried copper centers (one type-2; T2 and two type-3; T3; T2 formed by one ion and, together with the T3 group, which contains two ions) involved in oxygen reduction and water release (Georgiou et al. 2014; Kumar and Chandra 2020d). The substrates oxidized by the Lac lose electrons that are transferred to O_2 and the enzymatic electron transport system occurs from T1 to T2 and T3. The O_2 is then reduced to H_2O at the Cu sites T2 and T3, while the substrates usually give free radicals that may participate in other reactions, such as polymerization and hydration. For these reasons, the Lac can be used for the detoxification of several industrial effluents. Figure 6.1 shows the three-dimensional ribbon structure of bacterial laccase from *Bacillus subtilis* XI based on the homology modeling of *B. subtilis* MB24 laccase PDB Id: 2x88A (Guan et al. 2014) and fungal laccase from *Trametes versicolor*, respectively (PDB Id: 1N68) (Robert et al. 2003).

Microorganisms oxidize a wide range of environmental pollutants such as polyamines, aminophenols, lignin, aryl diamine, inorganic ions, and polycyclic hydrocarbon as a sole carbon or nitrogen source for their growth and metabolism and it may mitigate the toxicity of some compounds. Lac has a broad range of substrate specificity and the typical substrate of Lac known to be diphenol oxidase, 2,20-azino-bis(3-ethylbenzothiazoline-6-sulfonic acid) (ABTS), and monophenol, e.g., sinapic acid or guaiacol. The guaiacol is also a chromogenic substrate that is used for quick screening of microbial strains producing laccases by means of a color reaction. The selected four bacterial strains produced the laccase that catalyzed the oxidation of guaiacol to form reddish brown halo zones in the medium, as illustrated in Figure 6.1. During wastewater treatment, Lac catalyzes the oxidation of various phenolics (*ortho*- and *para*-diphenols, aminophenols, and polyphenols) and their derivatives, such

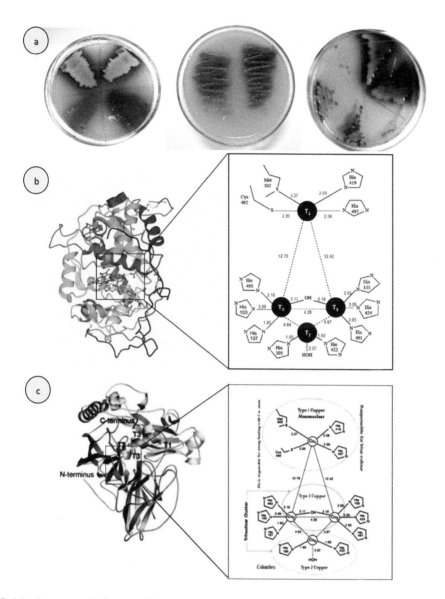

FIGURE 6.1 Laccase-producing bacteria and three-dimensional ribbon diagram of laccase (a) The oxidation of guaiacol to form reddish brown halo zones in the medium (b) The overall three-dimensional ribbon structure of bacterial laccase from *B. subtilis* and its catalytic centers based on the homology modeling of *B. subtilis* MB24 laccase PDB Id: 2x88A) (c) The three-dimensional ribbon structure of fungal laccase from *Trametes versicolor* and its two catalytic centers.

as ethers, aromatic compounds, and other non-phenolic compounds as well as some inorganic ions, particularly those with electron-donating groups such as phenols (OH) and anilines (NH_2), reducing oxygen as an electron acceptor to water by removing an electron from the phenols and aromatic amines, producing radicals. These radicals can undergo further Lac-catalyzed reaction and/or non-enzymatic reaction such as polymerization and hydrogen abstraction. The phenolic substrate oxidation by Lac results in the formation of aryloxy radicals and active species that is converted to a quinone in the second stage of oxidation. The biological treatment of distillery effluent usually requires enzymes to remain active under high pH and temperature conditions. Junior et al. (2020) evaluated the simultaneous treatment of raw vinasse, the Lac production by *Pleurotus sajor-caju*, and its purification using aqueous two-phase systems (ATPS), a low cost system that guarantees excellent quality levels of purity and enzymatic activity. ATPS is formed when two water-soluble compounds are mixed above its critical concentrations, resulting in two immiscible phases. Scientists have observed that the higher Lac production promoted a fourfold increase by using 0.4 mM of $CuSO_4$ as inducer, with the maximum enzymatic activity of 539.3 U/L on the third day of fermentation. The final treated vinasse had a decolorization of 92% and turbidity removal of 99% using $CuSO_4$. Moreover, the produced laccase was then purified by ATPS in a single purification step, reaching 2.9-fold and recovered 99.9%, in the top phase (PEG-rich phase) using 12 wt% of PEG 1500 + 20 wt% of citrate buffer + enzyme broth + water at 25 °C. The presence of lignocellulosic material can induce the production of MnP and laccase, both described as having a synergic action in the degradation of lignin and industrial effluents. Tapia-Tussell et al. (2015) studied the expression of Lac genes in the *Trametes hirsuta* Bm-2, isolated in Yucatan, Mexico, in the presence of phenolic compounds, as well as its effectiveness in removing colorants from vinasse. In the presence of all phenolic compounds such as guaiacol, ferulic acid, and vanillic acid, increased levels of Lac-encoding mRNA were observed compared to control. The lcc1 and lcc2 genes of *T. hirsuta* Bm-2 were differentially expressed; guaiacol and vanillin induced the expression of both genes, whereas ferulic acid only induced the expression of lcc2. The discoloration of vinasse was concomitant with the increase in Lac activity. The highest value of enzyme activity (2543.7 U/mL) was obtained in 10% (v/v) vinasse, which corresponded to a 69.2% increase in discoloration. Enzymatic decolorization of molasses medium has also been tried using *Phanearochaete chrysosporium*. Under stationary cultivation conditions, none of the strains decolorized molasses nor produced enzymes like LiP, MnP, and Lac. All of them could produce LIP and MnP when cultivated in flat bottom glass bottles under stationary cultivation conditions. According to Rodrıguez et al. (2003), the presence of phenolic compounds and flavonoid precursors, in the case of vinasse (known as laccase inducers), could favor the synthesis of this enzyme and a continuous activity for a longer period. *P. sajor-caju* demonstrated a rise in biomass production (1.06 g/100 mL), and the enzyme activities such as Lac (varying from 400 to 450 U/L) reached between the 9th and 10th day of growth and for MnP on the 12th day of cultivation (varying from 60 to 100 U/L). Authors concluded that the system *P. sajor-caju*/vinasse can be utilized as a bioprocess for color removal and degradation of complex vinasse compounds. González et al. (2008) mentioned that melanoidin-containing wastewaters induce expression of laccase gene (lcc1 and lcc2) in *Trametes* sp. I-62, a white-rot fungus belong to basidiomycetous fungi. The author also reports that rapid induction of lcc genes in media with complete molasses effluent may be associated with the presence of low-molecular-weight compounds that can be easily transported through fungal membranes. The inductive effect of melanoidins at the genetic level may be mediated by the action of lower molecular weight compounds derived from their degradation. Another factor to consider is that melanoidins are potent copper chelators. It has been shown that their chromophore groups are related to this property since liberation of chelated copper was detected when melanoidin degradation occurs as part of the decolorization process. If we consider that laccase genes can be induced by copper, then degradation of melanoidins could result in the induction of lcc gene expression as a consequence of the release of copper into the culture media. It was observed as improvement in the characteristics and detoxification allowing its utilization as reused water, Lac and MnP enzymes production, and for fungal biomass production with a high nutritional value (Ferreira et al., 2010). Pant and Adholeya (2009) mentioned the concentration of fungal ligninolytic enzymes by ultrafiltration and their use in distillery effluent decolorization. The maximum decolorization (37%) was obtained using the enzyme extract of *P. florida* EM1303. These concentrated enzymes when used for effluent treatment are capable of substantially decolorizing even the undiluted distillery effluent. Maximum per unit laccase (14.44 U/g) and MnP production (142.2 U/g) were observed in *Fusarium verticillioides* TERIDB16, while maximum LiP production (137.42 U/g) was in *Alternaria gaisen* TERIDB6. Fungal treatment of vinasses with WRF *T. versicolor* was performed to evaluate the fungus potential to produce laccase and decrease the concentration of phenol and chromophoric compounds in vinasse (España-Gamboa et al. 2015, 2017). Achieving 60% removal of COD, and over 80% of total phenol, with a decrease in almost 20% in color, *T. versicolor* has proven to be excellent laccase-producing microorganisms, achieving production of 1630 laccase units per liter of medium (España-Gamboa et al. 2015).

6.2.2 Manganese Peroxidases and Their Role in Distillery Effluent Treatment

MnP (E.C. 1.11.1.13. Mn^{2+}:H_2O_2 oxidoreductases) belongs to the family of oxidoreductases, to be specifically those actions on peroxide as acceptor (peroxidases), is an extracellular hemeprotein secreted in multiple isoforms under carbon- and nitrogen-limited media, which catalyze the H_2O_2-dependent oxidation of lignin or lignin-derivatives-based polymers in the presence of manganese ions (Mn^{2+}). MnP was first discovered in the mid-1980s in white-rot fungus *P. chrysosporium* by two international research teams (Gold's M and Crawford's R groups) and characterized as another key oxidative enzyme for lignin degradation. After nearly simultaneous discovery, it has been reported in a large number of ligninolytic fungi, including *Phlebia radiata, Pleurotus ostreatus Bjerkandera adusta, Dichomitus squalens, T. versicolor, Lentinus edodes*, and so on. It is produced by almost all wood colonizing white root and several litter decomposing basidiomycetes during secondary metabolism in response to nitrogen or carbon starvation. MnP has also been produced by some native bacterial strains like *Klebsiella pneumoniae* IITRCS01, *Salmonella enterica* IITRCS06, *Enterobacter aerogenes* IITRCS07, and *Enterobacter cloaceae* IITRCS11 (Kumar and Chandra 2018a); *Klebsiella* sp. (B2–B3), *Escherichia coli* (B4), and *Lactobacillus kefiri* (B1) (Omar et al. 2021); *Proteus mirabilis* IITRM5, *Bacillus* sp. IITRM7, *Raoultella planticola* IITRM15, and *Enterobacter sakazakii* IITRM16 (Yadav and Chandra 2012); *Bacillus licheniformis* (RNBS1), *Bacillus* sp. (RNBS3), and *Alcaligenes* sp. (RNBS4) (Bharagava et al. 2009); and *Bacillus* sp. IITRM7, *Raoultella planticola* IITRM15, and *Enterobacter sakazakii* IITRM16 (Yadav et al. 2011). MnP is glycoproteins having a molecular weight ranging from 32 to 62.5 kDa, ~350 amino acid residues at optimum pH of 4–7, and optimum temperature of 40–60°C; it showed 43% identity with LiP sequences. It is a specific enzyme that can oxidize Mn^{2+} to Mn^{3+}, which diffuses from the enzyme surface and in turn oxidizes the phenolic substrate, including lignin model compounds and some organic pollutants. In nature, MnP catalyzes plant lignin depolymerization due to its high potential for penetrating deep into the soil fines as a component of ligninolytic enzymes complex. So it is one of the most common lignin-degrading enzymes and has great application potential in the field of agriculture for degradation of lignocellulosic waste (Kumar and Chandra 2020d). To protect the environment, it was widely used in many industrial fields for degradation of some recalcitrant organic pollutants, such as simple phenol, amines, dyes, phenolic lignin substructure and dimers, and veratryl alcohol (VA: 3,4-dimethoxybenzyl), which are very harmful to human and animal health. The phenol red changed from deep orange to light yellow color during screening of peroxidase activity in bacteria that has been used as an indicator of MnP activity shown in Figure 6.2. This change in color of phenol red occurred due to oxidation of glucose by sugar oxidase enzyme, resulting in the production of H_2O_2 and media acidification (i.e., lowering of pH), which is required for the melanoidins degradation.

The protein molecule of MnP contains ten major helices and one minor helix. The molecular structure of MnP and its Mn^{2+} binding site is shown in Figure 6.2. MnP catalyzes the oxidation of Mn^{2+} to Mn^{3+} chelate with organic acid to form stable complexes that diffuse freely and oxidized various phenolic and non-phenolic substrates. Therefore, a wide range of substrate-oxidizing capability renders it an interesting enzyme for biotechnological applications in several industries. MnP catalyzes the oxidation of Mn^{2+} to Mn^{3+} in the presence of 1 equiv H_2O_2 as a co-substrate. At the first step of the catalytic cycle, MnP forms MnP-I, a high-valent oxo-Fe^{4+} porphyrin-based (Pi) free radical cation (step 1), which is in turn reduced by a bound Mn^{2+} atom to form MnP-II, an oxo-Fe^{4+} porphyrin without the associated porphyrin (Pi) free radical (step 2). However, in the absence of Mn^{2+}, the addition of 2 equiv H_2O_2 yields MnP-II. The conversion of MnP-I to MnP-II can also be achieved by the addition of other electron donors. Afterward, MnP-II oxidizes another Mn^{2+} ion, driving the enzyme back to the ground state Fe^{3+} porphyrin (step 3). In the absence of substrate, the addition of excess H_2O_2 (250 equiv) drives MnP into compound-III (MnP-III), which can be further oxidized until bleaching and irreversible inactivation (steps 4 and 5) (Figure 6.2). The Mn^{3+} is a strong oxidizer (1.54 V) and released from the MnP, but it is quite unstable in aqueous media. To overcome this drawback, WRF secrete various organic acids that act as chelating agents, enabling the formation of organic acid-Mn^{3+}complex. The complex formation stabilizes Mn^{3+} so that bidentate-ligated Mn^{3+} usually have redox potentials of around 0.7–0.9 V and significantly lower oxidation capacities when compared to

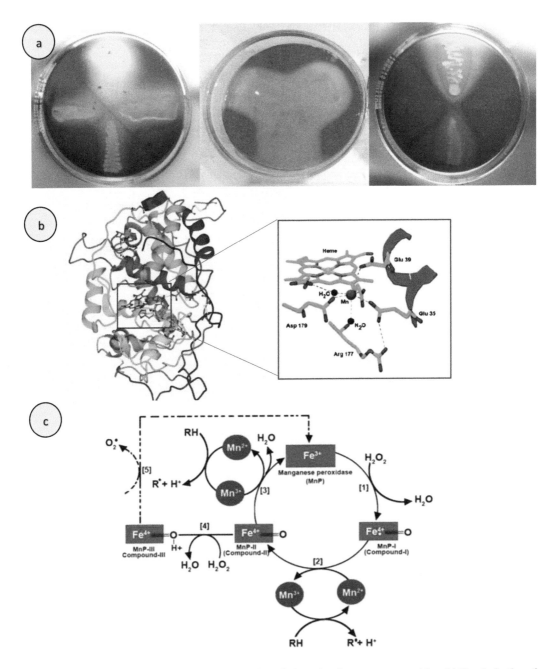

FIGURE 6.2 Three-dimensional ribbon diagram and catalytic cycle of manganese peroxidase (a) Decolorization of phenol red amended media due to production of manganese peroxidase (b) The overall three-dimensional ribbon diagram of microbial manganese peroxidase (c) Catalytic cycle of manganese peroxidase (Chandra et al. 2017).

non-chelated Mn^{3+}. The redox potential of chelated Mn^{3+} depends on the chelator. The degradation of recalcitrant non-phenolic compounds has been limited with MnP-generated Mn^{3+} chelates alone due to this lower oxidation power, but in the presence of some mediators or co-oxidants such as glutathione, polyoxyethylene sorbitan monoleate (tween 80), acetosyringone, methyl syringate, 3,5-dimethoxy-4-hydroxy-benzonitrile, linoleic acid, and linolenic acids, it is effective in the oxidation of recalcitrant compounds. There are various phenolic compounds—e.g., ABTS, 2,6-dimethyloxyphenol (DMP), vanillylacetone, ferulic acid, syringol, guaiacol, isoeugenol, *p*-methoxyphenol, syringaldazine, divanillylacetone, phenol red, and coniferyl alcohol, 3-(dimethylamino) benzoic acid (DMAB), *p*-cresol, *o*-dianisidine, catechol—and non-phenolic compound—e.g., vanillyl alcohol, VA, and benzyl alcohol—used for in vitro MnP assay. Among them, the widely used substrate for MnP assay is guaiacol. Several bacteria and fungi used for the decolorization of distillery effluent comprise MnP and Lac. Miyata et al. (1998) elucidated that extracellular H_2O_2 and manganese-independent peroxidase (MIP) and MnP have been involved in melanoidin decolorization by *Coriolus hirsutus*. Both enzymes were considered to be key enzymes in the decolorization. *Candida glabrata* yeasts are isolated from the mixed liquor of an activated sludge reactor and produced MnP able to degrade and decolorize melanoidins solution by co-metabolism (Yang et al. 2008). Microbial decolorization of melanoidin-containing wastewaters with combined use of activated sludge and a WRF, *C. hirsutus*, was studied by Miyata et al. (2000). *C. hirsutus* exhibited a strong ability to decolorize melanoidin in cultures not supplemented with nitrogenous nutrients. For enhancing the decolorization of melanoidin in wastewaters by the fungus, activated sludge pretreatment of the wastewaters was expected to be effective, i.e., activated sludge is capable of converting available organic N into inorganic N. Pretreatment of waste sludge heat treatment liquor (HTL) under appropriate conditions accelerated the fungal decolorization of HTL. Miyata et al. (2000) observed that the addition of Mn^{2+} to the pretreated HTL caused a further increase in the decolorization efficiency of *C. hirsutus* and a marked increase in the MnP activity. Taskin et al. (2016) reported the decolorization of molasses melanoidins by cold-adapted filamentous fungus *Cladosporium herbarum* ER-25 in non-sterile molasses medium. *C. herbarum* ER-25 removed melanoidins in non-sterilized molasses medium through biodegradation and bioadsorption mechanisms. A color removal rate of 64.8% in non-sterilized molasses medium could be achieved at the end of 144-h cultivation period. They observed that Lac and MnP were responsible for biodegradation and decolorization of melanoidins and molasses, respectively.

6.2.3 Lignin Peroxidases and Their Role in Distillery Effluent Treatment

LiP (EC 1.11.1.14) is a glycoprotein that has a prosthetic group made up of iron protoporphyrin IX, with catalytic activity dependent on H_2O_2. The H_2O_2 required for LiP activity originates from different biochemical pathways, expressed according to the nutritional factors and growth conditions of the microorganisms. LiPs were first discovered in *P. chrysosporium* (Tien and Kirk 1983; Paszczynski, Huynh and Crawford 1986) and later in *T. versicolor* (Johansson and Nyman 1993), *Bjerkandera* sp., and *Phlebia tremellosa*, which are well-known WRF The activity of LiP has previously been detected in some bacteria, such as *Acinetobacter calcoaceticus* and *Streptomyces viridosporus* (Dashtban et al. 2010). LiP catalyzes oxidations of phenolic and non-phenolic substrates generating phenoxy and aryl cation radicals cleaving side chains of these compounds and catalyzing aromatic ring-opening reactions, demetoxilation, and oxidative dechlorination. LiPs are usually secreted by microorganisms as a family of isozymes whose relative composition and isoelectric points (pI) vary depending on the growth medium and nutrient conditions. The globular structure of LiP isolated from *P. chrysosporium* is composed of eight major and eight minor α-helices with limited β-components and is organized into two domains that form an active center cavity composed of a heme-chelating single ferric ion. LiP contains two glycosylation sites, two Ca^{2+} binding sites, and four disulfide bridges, all stabilizing the three-dimensional structure of this enzyme. The molecular mass of LiPs from different WRF strains varies from 37 to 50 kDa and it has a pI between 3.1 and 4.7. pH and temperature activity profiles of LiPs from different sources vary significantly with optimum activities shown between pH 2–5 and temperature 35–55 °C. The high redox potential of LiPs (around 1.2 V at pH 3) makes these enzymes capable of oxidizing substrates that are not oxidized by other peroxidases. The catalytic cycle of LiP resembles the catalytic mechanism common to all peroxidases. In each cycle, the native enzyme is oxidized by H_2O_2 and generates a compound I intermediate that exists as a ferry oxo-porphyrin radical cation [Fe(IV) = O⁺]. Next, the enzyme is subjected to two single-electron reduction steps by the electron donor substrate, such as VA, leading to a transient formation of compound II [Fe(IV) = O] and a VA radical cation (VA⁺). The compound II further oxidizes the second VA molecule, simultaneously returning to its native stage to initiate a new catalytic cycle of LiP. VA, similar to Mn^{3+}, plays the role of being a small-molecular-weight redox transfer mediator between the enzyme and its polymeric substrate (Wong 2009). LiP possesses a catalytically active tryptophan residue exposed on the enzyme surface.

Typical oxidation reactions of LiP include Cα-Cβ cleavage, aromatic ring cleavage, and demeth(ox)ylation. Immobilization of LiP has been found to enhance its pH and temperature optima as well as thermostability and catalytic properties. The natural fungal secondary metabolites VA and 2-chloro-1,4-dimethoxybenzene act as redox mediators to stimulate the LiP-catalyzed oxidation of a wide range of recalcitrant substrates (Figure 6.3).

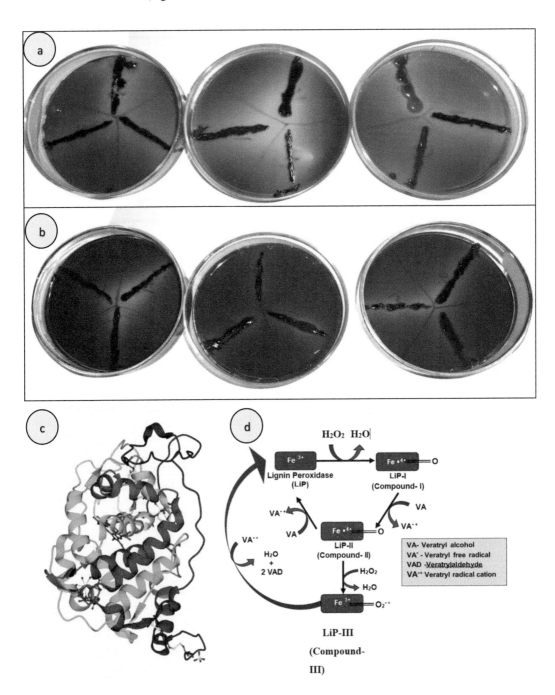

FIGURE 6.3 Lignin peroxidase producing bacterial strains, its catalytic cycle, and three-dimensional ribbon diagram. (a) The oxidation of methylene blue to form whitish zones in the medium (b) The oxidation of azure B to form whitish zones in the medium (c) The overall three-dimensional ribbon structure of lignin peroxidase (d) Catalytic cycle of lignin peroxidase.

6.2.4 Versatile Peroxidases and Their Role in Distillery Effluent Treatment

Versatile peroxidases (VP) (EC 1.11.1.16), also known as hybrid peroxidase (manganese-lignin peroxidase), a recently described family of ligninolytic peroxidases, show a hybrid molecular architecture of the MnP and LiP fungal peroxidase families combining different oxidation sites connected to the heme cofactor. VP is characterized by a broad substrate preference and by sharing typical features. VP was detected in members of the genera *Pleurotus eryngii*, *P. ostreatus*, *Bjerkandera adusta*, and *B. fumosa*. The production of peroxidase is shown by various fungal species such as *Bjerkandera, Lipista, Pleurotus*, and some bacterial species (VPs are glycoproteins secreted as several isoenzymes with molecular mass ranging between 40 and 45 kDa and pI ranging between 3.4 and 3.9. The structure of VP includes 11–12 helices, four disulfide bridges, two structural Ca^{2+} sites, a heme pocket, and an Mn^{2+} binding site similar to that of MnP. The substrate specificity of VP is similar to that of LiPs, including oxidation of high and medium redox potential compounds. These groups of enzymes are capable for the oxidation of phenolic, non-phenolic, and lignin derivatives in the absence of manganese. VP also oxidizes azo-dyes and other non-phenolic compounds with high-redox potential in the absence of mediators. They did not require any mediator for oxidation of compounds. The basic catalytic cycle of VP is similar to those of other peroxidases with the two intermediary compounds I and II, but is more complex due to a more diversified pool of potential substrates.

6.3 Ligninolytic Enzyme Immobilization

The industrial application of free ligninolytic enzymes is restricted due to their high price and instability and lack of reusability in aqueous solution when exposed to high temperatures, extreme pH conditions, organic solvents, or toxic reagents. Due to the versatile industrial and other applications of ligninolytic enzymes, the biotechnologists have used different types of solid supports and gels for developing effective catalysts. Immobilization of enzymes stabilizes catalysts against pH and temperature shifts, inhibitors, denaturants, and organic solvents and practical and chemical denaturation and catalytic activity of the enzyme for repeated use in effluent decolorization. Enzyme immobilization, first commercialized in the 1960s, has been developed as a unique chemical-based engineering technique that facilitates the reusability and recovery of enzymes. Importantly, high stabilization of an enzyme is often achieved through immobilization, which can trim-down the cost of enzyme-based industrial catalysis. Ligninolytic enzymes have been immobilized by using diversified techniques under particular conditions to fetch an improved yield for upgrading industrial benefits and enable the reusability of biocatalysts and impact on a lot of parameters like general catalytic efficiencies, the efficacy of catalyst utilization, thermostabilities, deactivation, reusability and regeneration of kinetics, and cost in most of the investigated cases concerning free-state of the enzyme. A range of novel carriers and solid-supporting matrices like alginate/carbon composite beads, mesostructured cellular foams, magnetic bimodal mesoporous carbon, magnetic chitosan beads, nanoporous silica on magnetic chitosan-clay composite beads, alumina pellets, composite hydrogels, and carbon supports have been utilized for the immobilization of ligninolytic enzymes from different sources. Moreover, different techniques have also been used for laccase immobilization adsorption to insoluble materials, encapsulation in membranes, cross-linking with bifunctional reagent, or covalent linking to an insoluble carrier, entrapment in polymeric gels, or a bioinspired immobilization entrapment technique for formation of laccase-biotitania biocatalysts suitable for environmental applications. Chavan et al. (2013) provided a reasonable alternative for cost-effective bioremediation of distillery spent wash using immobilized *Aspergillus oryzae* on baggase fibers. Authors treated melanoidin-containing distillery spent wash effluent by free and immobilized *A. oryzae* MTCC7691. Fungal mycelia immobilized on baggase packed in a glass column under a batch-wise mode (i) effected in removal of 75.71% color, 51.0% BOD, 86.19% COD, and 49.0% phenolic pigments of distillery spent wash up to 25 days at 30 °C, while free fungal mycelia resulted in removal of 63.1% color, 27.74% BOD, 76.21% COD, and 37.32% phenolic pigments of distillery spent wash using shake flask; (ii) MnP activity was highest (1.55 ± 0.01 U/mL/min) in immobilized fungi, followed by LiP (0.65 U/mL/min) and Lac activity (0.9 CU mL/min); (iii) accumulative MnP activity was highly correlated with spent wash decolorization and reduction of phenolic pigments, suggesting the presence of MnP activities in bioremediation of spent wash; and (iv) degradation of spent wash was confirmed by high-performance thin-layer chromatography and gas chromatography-mass spectrometry analysis. Pant and Adholeya (2007a) analyzed the decolorization of distillery effluent using immobilized fungal biomass of *A. niger* TERI DB20, *F. verticillioides* ITCC, and *A. flavus* TERI DB9 at a higher concentration (50% v/v). Authors achieved maximum decolorization (86.33%) in *P. ostreatus* (Florida) Eger EM 1303 immobilized on corncob with molasses in 28 days. Singh et al. (2015) used laccase immobilized on alumina pellets activated with aminopropyltriethoxysilane (APTES) for decolorization of glucose-glycine Maillard products. The immobilization yield was 50–60%, and the enzyme activity (886 U/L) was fivefold higher compared to the soluble enzyme (176 U/L). The immobilized enzyme also showed higher tolerance to pH (4–6) and temperature (35–60 °C), as well as improved storage stability (49 days) and operational stability (10 cycles). Degradation of glucose-glycine Maillard products using immobilized Lac led to 47% decolorization in 6 h at pH 4.5 and 28 °C. A comprehensive treatment scheme integrating enzymatic, microbial, and membrane filtration steps resulted in 90% decolorization. Georgiou et al. (2014) have reported the immobilization of laccase from *T. versicolor*, performed on alumina beads or on controlled pore glass-uncoated particles, for decolorization of melanoidin from baker's yeast effluents. Georgiou et al. (2016) studied the decolorization of melanoidins from simulated and industrial molasses effluents by immobilized Lac. The immobilized enzyme showed higher tolerance to pH (4–6.5) and temperature (35–60 °C) and improved thermostability maintaining at 80 °C. Immobilized Lac displayed operational stability (11 cycles for syringaldazine and 4 cycles for melanoidin). Degradation of simulated molasses wastewaters after 48 h with immobilized Lac on glass and alumina reached 74% and 71%, respectively. Whereas degradation of baker's yeast effluents by immobilized Lac on glass reached 68% within 24 h at pH 4.5 and 28 °C for a melanoidin solution 1% v/v.

6.4 Miscellaneous Enzymes in Degradation and Decolorization of Melanoidins-Containing Effluent

Several reports claimed that the intracellular sugar-oxidase-type enzymes (i.e., sorbose-oxidase or glucose oxidase) had melanoidin-decolorizing activities. It was suggested that melanoidins were decolorized by the active oxygen ($O_2:H_2O_2$) produced by the reaction with sugar oxidases. *Coriolus* sp. No. 20 decolorized a melanoidin solution, a decrease of about 80% in darkness under the optimal conditions (pH 4.5 and temperature 35 °C) (Watanabe et al. 1982). This decolorization occurred with an intracellular enzyme that was prepared from an extract of integrated mycelia, and required aeration and some kinds of sugars, particularly glucose and sorbose. The purified enzyme was identified as L-sorbose oxidase (L-sorbose: oxygen 5-oxidoreductase); decolorization proceeded in the presence of oxygen and sugars such as maltose, sucrose, lactose, galactose, and xylose, besides glucose and sorbose (Watanabe et al., 1982). However, glucose oxidase [β-D-glucose: oxygen I-oxidoreductase] also decolorized melanoidin pigments. Melanoidin was suggested to be decolorized by the active oxygen (O_2, H_2O_2) produced by the reactions with these oxidases because the reaction with the pure enzyme was accompanied by the oxidation of glucose to gluconic acid. Ohmomo et al. (1985) extracted and purified melanoidin-decolorizing enzymes, P-III and P-IV, from mycelia of *Coriolus versicolor* Ps4a. This strain decolorized molasses wastewater approximately 80% in darkness under the optimum conditions. Both enzymes were apt to decolorize molasses wastewater more than synthetic melanoidins. Among the tested synthetic melanoidins, glucose-glutamic acid melanoidin was decolorized strongly and the degree of decolorization was approximately 1.5 times higher than that of glucose-glycine melanoidin. Both enzymes also showed strong activity toward soybean sauce pigment as well as molasses wastewater, but hardly attacked corncob lignin. Kelley and Reddy (1986) identified the glucose oxidase activity as the primary source of H_2O_2 production in ligninolytic cultures of *P. chrysosporium*. Similarly, Chin Fa et al. (2011) purified and characterized a novel glucose oxidase-like melanoidin decolorizing enzyme

(MDE) from *Geotrichum* sp. No. 56. A novel MDE produced by *Geotrichum* sp. No. 56, which exhibits decolorization activity against synthetic melanoidin and molasses-containing wastewater, was purified and characterized. The purified MDE showed optimum activity at pH 6.5. The optimum temperature of the enzyme was 45 °C, above which the thermal stability decreased dramatically. Two fungal strains *Penicillium pinophilum* TERI DB1 and *Alternaria gaisen* TERI DB6 produced ligninolytic enzymes laccase, MnP, and LiP and have potential to decolorize distillery effluent that were isolated from the soil of a distillery effluent contaminated site. Reduction in color up to the magnitude of 86%, 50%, and 47% was observed with *P. florida, P. pinophilum*, and *A. gaisen*, respectively (Pant and Adholeya, 2007b). Decolorization of molasses spent wash using free and immobilized mycelia of *Flavodon flavus* is accompanied by simultaneous detoxification and decrease in polyaromatic hydrocarbon contents of the molasses spent wash possibly via the action of glucose oxidase, accompanied by the production of H_2O_2 that acts as a bleaching agent in the process (Raghukumar and Rivonkar 2001; Raghukumar et al. 2004). Decolorization and COD reduction (52%) of distillery spent wash by *C. versicolor* are dependent on the carbon source and addition of organic/inorganic nitrogen has no enhancing effect on decolorization and COD reduction (Chopra et al. 2004). Ahmed et al. (2018) studied the sustainable bioremediation of sugarcane vinasse using autochthonous macrofungi, i.e., *Pycnoporus* sp. P6 and *Trametes* sp. T3. The two basidiomycetous fungi were selected based on their ability to decolorize vinasse and to synthesize lignocellulolytic enzymes in agar medium. In liquid cultures, fungi demonstrated their capacity to both decolorize and remove phenolic compounds from diluted vinasse, with the concomitant production of ligninolytic enzymes, mainly laccases, suggesting the participation of this enzyme in the bioremediation process. Table 6.1 summaries the successful studies on decolorization of melanoid in containing distillery effluent.

TABLE 6.1

Numerous Successful Studies on Decolorization of Melanoidin-Containing Distillery Effluent by Immobilized Microorganisms or Their Extracellular Enzymes

Wastewater	Fungal Species	Ligninolytic Enzyme Activity			Decol (%)	Reference
		MnP	LiP	Laccase		
Colored effluents	An unidentified basidiomycete species NIOCC # 2a	ND	ND	2075 (U/L)	–	D'Souza et al. (2006)
Vinasses	*Pleurotus ostreatus*	–	–	14.1 (U/mL)	–	Rodríguez et al. (2003)
Molasses distillery wastewater	*Pleurotus ostreatus (Florida) Eger EM 1303, Aspergillus flavus TERI DB9, Fusarium verticillioides ITCC 6140, Aspergillus niger TERI DB18, A. niger TERI DB20*	0.34 U/mL (*F. verticillioides* ITCC 6140) 1.12 (U/mL) (*A. niger* TERI DB18) 0.81(U/mL) (*A. niger* TERI DB20)	2.03 U/mL for *F. verticillioides* ITCC 6140) 0.89 (U/mL) (*A. niger* TERI DB20)	0.45 (U/mL for *A. flavus* TERI DB9) 0.81(U/mL for *P. ostreatus* (Florida) EM1303	86.33, 74.67, 68.33, 64.67, 54.67	Pant and Adholeya (2007b)

(Continued)

TABLE 6.1 (*Continued*)

Wastewater	Fungal Species	Ligninolytic Enzyme Activity			Decol (%)	Reference
		MnP	LIP	Laccase		
Distillery effluent	*Fusarium verticillioides* TERIDB16, *Alternaria gaisen* TERIDB6 and *Pleurotus florida* EM1303	142.2 U/g for *F. verticillioides* TERIDB16	137.42 U/g for *Alternaria gaisen* TERIDB6	14.44 U/g for *Fusarium verticillioides* TERIDB16	37	Pant and Adholeya (2009)
Melanoidin solution	*Coriolus* sp. No. 20	Sorbose oxidase	–	–	80	Watanabe et al. (1982)
Melanoidin wastewater	*Candida glabrata* Y1	Detected	–	–	70	Mahgoub et al. (2016)
Simulated and industrial molasses effluents	*Trametes versicolor* (53739-1G-F)	–	–	Immobilized	74 and 71	Georgiou et al. (2016)

ND: not detected.

7

Bioreactors in Distillery Wastewater Treatment

7.1 Introduction

Increased power requirements and the need to preserve the environment have motivated the search for renewable energy sources. As one alternative, several countries have promoted bioethanol production programs using sugar-based raw materials. The distillery industry is considered as one of the water-intensive industries that produce a high volume of wastewater. In particular, distilleries discharge significant quantities of recalcitrant organic compounds, heavy metals, melanoidins (result from the reaction of sugars and proteins by the Maillard reaction), phenolics, endocrine-disrupting chemicals, and various androgenic-mutagenic compounds, which ultimately results in high turbidity, color, biological oxygen demand (BOD), chemical oxygen demand (COD), dissolved solids, suspended solids, and organic matter in the receiving water environment. The composition and concentration of effluent depend on both the raw material used for sugar extraction and the subsequent fermentation and distillation processes. Free discharge without proper treatment can inflict large, adverse impacts on the environment. To date, a significant number of distillery effluent treatment processes have been reported in the literature. Some of the well proven methods are photo-electrochemical process, adsorption, ion exchange, membrane filtration, chemical precipitation, flotation, advanced oxidation processes, coagulation/flocculation, and biological treatment using fungi, bacteria, algae, and biocoagulants extracted from plants, Most of these processes either convert the toxic substances into non-toxic substances or separate from the aqueous phase. The treatment of distillery effluent in lagoons generates greenhouse gas emissions, and ferti-irrigation practices may in some cases negatively affect the structure of soils and aquifers. Spent wash also takes a high value as fertilizer, due to its high organic matter, and micronutrients content is often used in crops fertigation. However, when used in large quantities, spent wash can saturate the soil and contaminate nearby water bodies. On the other hand, due to its very high concentration of organic matter, spent wash could be a potential resource for energy recovery, producing biogas via anaerobic processes. However, the major disadvantage of anaerobic digestion is the strong and recalcitrant color of the anaerobically digested stillage. The presence of phenolic compounds (8000–10,000 mg/L), melanoidins, caramel, and the furfural components contributes to its color and makes spent wash complex and difficult effluent for degradation. Such recalcitrant color is of great concern because some colored compounds are not only recalcitrant to biodegradation but also inhibitory to microbial activities in the biological treatment system. Therefore, the design of effective treatments to remove color from untreated, partly treated, anaerobically digested distillery effluent is of great interest. This chapter focuses on the working principle of different types of bioreactors, particularly those used in the treatment of distillery spent wash. The types of bioreactors dealt with in this chapter include aerobic bioreactors, anaerobic bioreactors, plug flow bioreactors, upflow anaerobic sludge blanket bioreactors, photobioreactors, reverse membrane bioreactors, immersed membrane bioreactors, fluidized bed bioreactors, packed bed bioreactors, activated sludge bioreactors, membrane bioreactors, and immobilized cell bioreactors. Mostly these bioreactors are designed in such a way that they can be engineered or modified to be used for sewage or wastewater treatment and the production of biodiesel, bioethanol, biogas, etc. The overall objective of this chapter is to summarize the different types of bioreactors and their role in degradation and decolorization of wastewater generated from distilleries.

DOI: 10.1201/9781003029885-7

7.2 What Is Bioreactor and Their Principles?

Bioreactors are confined space environments used to promote microbial growth and to perform fermentative processes under controlled physicochemical parameters such as agitation, temperature, pH, aeration, and feeding. They are used to promote the synthesis of many bioproducts such as enzymes, antibiotics, and organic acids, as well as applied in the treatment of industrial waste. An anaerobic treatment of spent wash has often been cited as an effective and economical treatment option because it eliminates the COD and converts it to biogas, which is a readily usable fuel for the ethanol facility. These advantages of anaerobic treatment result in a reduction of the operational costs compared with conventional wastewater treatment. Although anaerobic digestion of most types of vinasses is feasible and quite appealing from an energy point of view, the presence of recalcitrant compounds can be toxic or inhibitory for anaerobic microorganisms, commonly phenols, melanoidins, and a variety of sugar decomposition products. This slows down the kinetics and reduces mean rates of methane production and yield coefficients. Moreover, the presence of phenolic compounds in wastewater represents a health and environmental hazard. There are many high-rate anaerobic reactors that have been used for high-strength organic wastewater treatment. Some of them like expanded granular sludge bed (EGSB), hybrid reactors, granular-bed anaerobic baffled reactor (GBABR), upflow blanket filter internal circulation reactor (ICR), fixed bed reactor, upflow anaerobic filter, and fluidized bed reactor (FBR) have been studied for the treatment of wastewater by different researchers. Each of the reactors has their own limitations such as usefulness at lower temperatures only (4–20°C) and treatment of relatively very low-strength wastewaters, and inadequacy for the removal of particulate organic for EGSB. Cost, maintenance, and more monitoring problems are associated with hybrid reactors. Biofilm formation on carriers poses problems leading to long start-up times, difficulty in control of biofilm thickness, and clogging problem for ICR. More cost, careful monitoring requirement, channeling in reactor, service, and cleaning difficulties are problems in a fixed bed reactor. Foaming and flotation due to gas production, flooding problems, erosion by abrasion of particles, particle entrainment, higher initial capital costs, and erosion of internal components are drawbacks of fluidized bed reactor. Anaerobic treatment is perhaps the most economical method to destroy the pollutional load significantly. Some anaerobic processes involve use of media for immobilization of the bacterial mass, e.g., fluidized bed, media-packed filters, while some processes do not involve use of media at all as in case of the contact process, upflow sludge blanket, etc. In all these processes, fixed-film systems occupy a unique position because of its several exclusive advantages. One such system, plastic packed, stationary fixed film, downflow anaerobic filter is in operation treating spent wash generated from a cane-molasses-based distillery in India. The filter is designed as a single stage completely mixed reactor and does not require any pH adjustment of raw spent wash prior to feeding at steady state. The anaerobic treatment of a complex substrate, including suspended organic matter present in the spent wash, can be subdivided into four successive stages, each requiring its own characteristic group of microorganisms:

i. *Hydrolysis:* This stage involves the conversion of high-molecular-weight non-soluble organic compounds into soluble compounds, appropriate for use as a source of carbon and energy by growing microorganisms. Hydrolysis of organic matter is a rather slow process brought about by extracellular enzymes.

ii. *Acidogenesis:* This step involves the transformation of the soluble small organic molecules that were the result of the first stage into low-molecular-weight intermediate compounds.

iii. *Acetogenesis:* In this stage, lower chain volatile fatty acids produced during acidogenesis are utilized by a group of bacteria (acetogens) to produce acetate and carbon dioxide. Acid production results in the formation of acetic acid or case of instability, the higher fatty acids such as propionic, butyric, isobutyric, valeric, and isovaleric acid. The acid production rate is high as compared to the methane production rate, which means that a sudden increase in easily degradable (soluble) organics will result in increased acid production with subsequent accumulation of the acids.

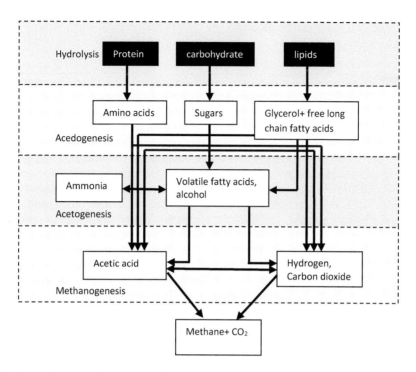

FIGURE 7.1 Schematic presentation of anaerobic digestion (biomethanation) process for waste treatment (Kumar and Sharma 2019).

iv. *Methanogenesis:* This stage involves the bacterial production of methane, primarily from acetate but also from hydrogen and carbon dioxide. Methane production is a slow process, in general the rate-limiting step of anaerobic degradation. About one-third of the methane has its origin in molecular hydrogen. Small amounts of methane can be produced from methanol and formic acid, but these reactions have little practical importance. The biochemistry and microbiology of anaerobic processes is much more complicated than that of aerobic ones. This is a result of the many pathways available for an anaerobic community. A general outline of the biomethanation process of waste treatment and biogas production is illustrated in Figure 7.1. In the anaerobic digestion process, any organic compound present in the spent wash is digested (metabolized) by the microorganisms in an oxygen-free environment to produce methane. Other products of the anaerobic digestion process are digested sludge and treated distillery effluent; both are highly rich in organic and inorganic nutrients. This anaerobically digested distillery sludge can be used as fertilizer.

Anaerobic digestion, also known as biomethanation, is the most attractive primary distillery spent wash treatment method because it is an eco-friendly, low-cost and socially acceptable microorganism-based technology (Mohana et al. 2007). The anaerobic methane fermentation process converts waste organic materials to methane and carbon dioxide in the absence of molecular oxygen. It has long been recognized as a useful process for wastewater treatment. The anerobic methane fermentation process offers several advantages over conventional aerobic treatment systems: (i) a higher degree of waste stabilization, (ii) a lower microbial yield, (iii) a lower nutrient requirement, (iv) no oxygen requirement, and (v) methane gas production. In the past, however, the broad-scale application of this process has been largely in the treatment of municipal sewage sludge and animal residues to achieve waste stabilization and solids reduction. Anaerobic digestion consumes one- to threefold the amount of freshwater for dilution to spent wash to ensure the appropriate condition of the anaerobic digester. The main challenges of anaerobic digestion of spent wash are the elimination of the color-contributing organic compounds and heavy metals. However, even after anaerobic digestion, the treated spent wash may have a high organic loading and a dark color, requiring additional treatment steps.

7.3 Commonly Used Bioreactors in Wastewater Treatment

7.3.1 Anaerobic Fluidized Bed Reactors

Anaerobic fluidized bed (AFB) is an anaerobic wastewater treatment method able to retain a high concentration of biomass with high specific activity and handle a high volumetric loading with efficient COD removal. It is particularly applicable to wastewater of lower concentration and wastewater containing biological inhibitive compounds under the high hydraulic loading operation. Reactor is often tall, less footprint is needed, inert carriers providing large surface area for attachment and growth of anaerobic microorganisms are added in the AFB bioreactor. The biofilm-attached carriers are fluidized through upflow water to enhance mass transfer and degradation of organic pollutants. Anaerobic treatment methods are more suitable for the treatment of concentrated wastewater streams, offer lower operating costs, and produce usable biogas product. AFB reactor is filled with fluidized carriers that provide large surface area for the growth of anaerobic or anoxic microorganisms. The input and recirculation water flow into the bottom of reactor and provide enough lift force for carriers fluidization. This upflow also delivers carbon sources effectively for microorganisms attached on carriers due to the known high mass transfer rate of fluidized bed reactor. The advantages of AFB are as follows: suitable for low concentration and bioinhibition of wastewater and can be operated in high hydraulic loading, high MLSS, and volumetric loading. A schematic view of anaerobic fluidized bed reactor is given in Figure 7.2.

7.3.2 Anaerobic Fixed Film Reactors

Another means of providing an anaerobic process with a high solids retention time for the methane-producing bacteria with a short hydraulic retention time for system economy is with fixed film reactors (Bachmann 1983). The fixed film reactors offer distinct advantages over other anaerobic systems such as simplicity of construction, elimination of mechanical mixing, better stability at higher loading rates, and capability to withstand large toxic shock loads. The solid support filling within reactor is the most important component of an anaerobic fixed film reactor. Some ideal characteristics of the packing material are high porosity, large surface area, adequate surface properties for adherence lightweight.

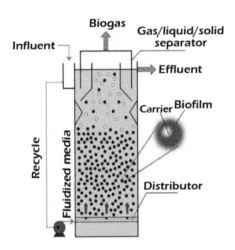

FIGURE 7.2 A schematic view of the anaerobic fluidized bed reactor.

In these heterogeneous systems, the microorganisms grow in a film on solid support while the organic matter is removed from the liquid flowing past them. These systems may have very high volume loading capacities compared with aerobic processes because they are not limited by oxygen transfer. Various types of fixed film reactor have been developed for anaerobic treatment. More cost, careful monitoring requirement, channeling in reactor, service, and cleaning difficulties are problems in a fixed bed reactor.

7.3.3 Expanded Granular Sludge Bed (EGSB) Reactors

An EGSB reactor is a variant of the upflow anaerobic sludge blanket digestion (UASB) concept for anaerobic wastewater treatment. The distinguishing feature is that a faster rate of upward-flow velocity is designed for the wastewater passing through the sludge bed. The EGSB bioreactor was developed in the Netherlands in the mid-1980s to increase AGS/wastewater contact and contribute to the reduction of the presence of dead zones, a preferential flow and short-circuits that can be carried out in the UASB bioreactor. In an EGSB reactor, the granular anaerobic sludge is fluidized due to the upward velocity of the liquid stream introduced by strong recirculation. The increased flux permits partial expansion (fluidization) of the granular sludge bed, improving wastewater-sludge contact as well as enhancing segregation of small inactive suspended particles from the sludge bed. The increased flow velocity is either accomplished by utilizing tall reactors or by incorporating an effluent recycle (or both). A schematic view of EGSB is shown in Figure 7.3. EGSB bioreactor has attracted many researchers, because it has several advantages like design simplicity, usage of unsophisticated equipment, low anaerobic granular sludge (AGS production, high treatment efficiency, low operating costs) and its potential to generate renewable energy (like biogas, biomethane, or biohydrogen) has turned this bioreactor into a sustainable alternative to mitigate the crisis of water pollution. A multiscale approach was followed, using batch tests, continuous bench scale reactors, and a full-scale UASB reactor. A load of 20 g COD/L per day with COD removal efficiencies between 60% and 80% were reached in benchscale UASB and EGSB reactors. Finally, a 100 m³ UASB reactor located in a distillery was tested under industrial conditions. A load of 0.6 kg COD/kg VSS per day was achieved with a COD removal efficiency of 78% (López et al. 2017). A scheme depicting the EGSB design concept is shown in this EGSB diagram. A full-scale study of high-rate anaerobic bioreactors for whiskey distillery wastewater treatment

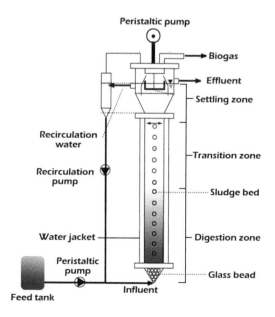

FIGURE 7.3 A schematic view of expanded granular sludge bed reactor.

with size fractionation and metagenomic analysis of granular sludge has been analyzed by Lin et al. (2020). Two full-scale high-rate bioreactors, i.e., external circulation sludge bed (ECSB) and EGSB, were monitored for three years. Their performances for treating wastewater in a whiskey distillery were compared in terms of COD, pH, alkalinity, and VFA. Even though feed flowrate highly fluctuated, COD removals of ECSB and EGSB were both excellent (95.7% and 94.8%, respectively). The influent and effluent characteristics of ECSB reactor were profiled and urea and urethane were also detected. High-strength properties of raw spent wash were exhibited in TOC, soluble COD and BOD_5, at 20 °C, of 13,500, 37,750, and 1950 mg/L, respectively, and characterized by gas chromatography-mass spectrometry (GC-MS). Anaerobic granular sludge sampled from different heights of ECSB reactor were fractionated for demonstrating vertical size distributions. Moreover, major species found by next-generation sequencing technique were archaea, i.e., *Methanosaeta* and *Methanolinea*, while major bacteria were *Bacteroidetes* with minor *Nitrospiraceae*. This metagenomic analysis provided an insight into anaerobic microbial consortium.

7.3.4 Upflow Anaerobic Sludge Blanket (UASB) Reactors

Upflow anaerobic sludge blanket technology, also known as UASB reactor, is a form of anaerobic digester that is used in wastewater treatment. UASB reactor was developed by Lettinga et al. (1980) whereby this system has been successful in treating a wide range of industrial effluents including those with inhibitory compounds. The underlying principle of the UASB operation is to have an anaerobic sludge that exhibits good settling properties (Lettinga, 1995) and efficiently retains complex microbial consortium without the need for immobilization on a carrier material (for example, as a biofilm) by formation of biological granules with good settling characteristics. Performance depends on the mean cell residence time and reactor volume depends on the hydraulic residence time; therefore, UASB reactor can efficiently convert organic compounds of wastewater into methane in small "high-rate" reactors. Approximately 60% of the thousands of anaerobic full-scale treatment facilities worldwide are now based on the UASB design concept, treating various range of industrial wastewaters (Latif et al., 2011). The UASB reactor is a methane-producing digester forming a blanket of granular sludge under an anaerobic condition to facilitate contact with the substrate; the sludge is processed by anaerobic microorganisms. UASB reactor is a methane-producing digester, which uses an anaerobic process and forms a blanket of granular sludge and is processed by the anaerobic microorganisms. UASB bioreactors entered the industrial scenario relatively late but quickly established their presence in the arena of anaerobic wastewater treatment due to its usefulness to continuously treat high-strength and non-diluted wastewaters. It can sustain higher organic loading rates (OLRs), and requires shorter hydraulic retention time (HRT) than other reactors. The UASB reactor exhibits positive features, such as high organic loading rates (OLRs), short hydraulic retention time (HRT), and a low energy demand. By contrast, in a UASB reactor, the sludge remains somewhat suspended, forming a "blanket" to facilitate contact with the substrate; this type of reactor also experiences lower upward velocity than in an EGSB reactor. A typical UASB system is shown in Figure 7.4.

The UASB has many advantages over other treatment units as given by Ma (2002); a gas-liquid-solid separator provided near the top of the reactor enables sludge to settle into the blanket, biogas to escape into the dome at the top of the reactor, and treated supernatant to flow out of the reactor. It is a compact unit, ideal for economic space utilization. The residues (sludge) generated by UASB are less in amount and are well digested compared to other anaerobic reactors, thus requiring reduced sludge handling and causing less odor problems. UASB reactor has no mechanical or moving parts to wear and tear. Thus, it is virtually maintenance-free and involves few operational problems. UASB treatment process requires no external input of energy. Biogas, rich in methane, is generated as a valuable by-product. Methane production is about 0.15–0.35 L/g COD destroyed. The required mixing is achieved by upflowing of wastewater and the rising gas bubbles. Both the capital cost and operating cost of a UASB reactor-based treatment plant are significantly less. The UASB reactor is a noiseless, closed, and covered unit that is esthetically very satisfying. Vlissidis and Zouboulis (1993) reported a full-scale thermophilic (50–55°C) anaerobic digestion of distillery effluent to achieve 60% COD removal with a recovery of 76% of biogas. Harada et al. (1996) carried out thermophilic (55 °C) AD

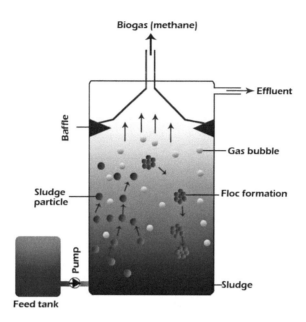

FIGURE 7.4 A schematic view of the upflow anaerobic sludge blanket reactor.

of DE using a 140-L UASB reactor for a period of 430 days. They achieved 39–67% COD and more than 80% removal of DE. The cheap performance of the reactor for COD elimination can probably be attributed to the low degradability of the waste itself. A two-stage mesophilic treatment system consisting of a UASB reactor and an anaerobic filter was found appropriate for anaerobic digestion of distillery effluent, enabling better conditions for the methanogens (Blonskaja et al. 2003). Harada et al. (2018) also worked with vinasse diluted to one-tenth in thermophilic UASB reactors and reported a significant decrease in methanogenic activity when using raw vinasse instead of acetate. They suggested that inhibition is associated with the presence of phenolic compounds. Goodwin et al. (2019) worked with vinasse from a whiskey distillery in UASB reactors and obtained good results only with diluted vinasse; otherwise, VFAs accumulated; beyond that, the pH did not decrease. Large hydraulic residence times contributed to accentuating these phenomena in the feeding area. They also indicated possible effects due to the high salinity of the effluent. Sharma and Singh (2001) investigated the effect of the addition of micronutrients and macronutrients in the DE treatment performance of the UASB system. Calcium and phosphate were found to be detrimental to treatment efficiency. Wolmarans and De Villiers (2002) have reported greater than 90% COD removal efficiency in a UASB reactor treating distillery effluent. The treatment of high-strength spent wash was studied in a benchscale UASB reactor (Saner et al. 2014). The reactor operated at mesophilic temperature (37 °C) at different OLRs and constant HRT in two-day periods over a research period of 635 days. The maximum BOD and COD removals achieved were 89.11% and 68.35%, respectively, at optimum OLR of 15.34 kg COD/m^3 day. Musee et al. (2007) indicated that the integration of the UASB reactor and an aerobic treatment process provides an improvement in COD removal from distillery effluent. An overall 96.5% COD removal was achieved through this hybrid (anaerobic/aerobic) effluent treatment. Saner et al. (2014) operated an UASB reactor at different OLRs and constant hydraulic retention time for 2 days at mesophilic temperature of 37°C for a period of about 2 years (635 days) for treatment of distillery wastewater. The maximum COD and BOD removals achieved were 68.35% and 89.11%, respectively, at optimum OLR of 15.34 kg COD/m^3 per day. UASB reactor performance was also evaluated in terms of hydrolysis, acidification, and methanogenesis, and the performance values were found to be 33.88%, 52.16%, and 48.07%, respectively. Total and soluble biodegradability of the high-strength wastewater were 48.09% and 78.06%, respectively, that represents the good conversion of soluble substrate to biogas. The average biogas produced was 0.38 m^3/kg COD removed. The COD mass balance of the reactor

TABLE 7.1

General Properties of Wastewater Used for UASB Process

Type of Wastewater	COD	BOD	TSS	VSS	pH	Reference
Distillery wastewater	100–120	30	51.5–100	2.8	3–4.1	Nataraj et al. (2006)
Vinasses	97.5 as SCOD	42.23	3.9	–	4.4	Martin et al. (2002)
Raw spent wash	37.5	–	2.82	–	4.2	Ramana et al. (2002)
Molasses wastewater	80.5	–	109	2.5	5.2	Jimenez and Borja (1997)
Alcohol distillery	11–33	6–16	–	–	4–7	Ince et al. (2005)

BOD: biological oxygen demand; COD: chemical oxygen demand; TSS: total suspended solids; VSS: volatile suspended solids.

showed that 51.32%, 0.24%, 9.46%, 1.75%, and 37.22% COD were converted into methane (gaseous phase), methane (aqueous phase), sludge, sulfate reduction, and effluent, respectively. UASB reactors are the most popular continuous reactor type used for the anaerobic treatment of wastewaters. However, the use of an EGSB reactors is increasing. Saini and Lohchab (2017) evaluated performance of UASB reactor to treat high-strength wastewater distillery spent wash under various operating conditions. The results showed COD reduction of 91%, 83%, and 72% in phases I, II, and III, respectively. BOD and VFA reduction were 89% and 78%, 84% and 80%, and 72% and 72% in phases I, II and III, respectively. **Table 7.1** summarizes the few reports on treatment of distillery wastewater by UASB system.

España-Gamboa et al. (2017) demonstrated the possibility of using a FBR for the elimination of recalcitrant compounds and COD at non-sterile conditions by *T. versicolor*. This shows that significant energy saving may be possible when the production media large volume (industrial application) are directly used without sterilization. Continuous operation of the fluidized bed bioreactor was carried out successfully for 26 days, which is the highest value found in the literature. Two systems were evaluated for the vinasse treatment in which System 2 coupled the FBR to the UASB reactor and registered the best quality of effluent and higher methane content in the biogas. Conclusively, the coupling of FBR to UASB reactor is a promising environmental technology for the treatment of vinasse; nevertheless, an economic study and feasibility of application of the process to full scale is necessary. Acharya et al. (2011) assessed microbial community of anaerobic biphasic fixed-film bioreactor treating high-strength distillery wastewater. Sequencing of 16S rRNA genes exhibited a total of 123 distinct operational taxonomic units (OTUs) comprising 49 from acidogenic reactor and 74 (28 of eubacteria and 46 of archaea) from methanogenic reactor. The findings reveal the role of *Lactobacillus* sp. (Firmicutes) as dominant acid-producing organisms in acidogenic reactor and *Methanoculleus* sp. (Euryarchaeotes) as foremost methanogens in the methanogenic reactor. The results show the applicability of Stover-Kincannon model to the anaerobic biphasic reactor that is efficient to remove 50–80% COD at different OLR. Phylogenetic analyses reveal the presence of acidogenic organisms (Firmicutes) in AR and dominance of hydrogenotrophic methanogens (Euryarchaeotes) in MR. Community profiling at every change in the HRT and OLR would be of great advantage in further understanding of the complex microbial process. Treatment efficiency of the bioreactor was investigated at different HRT and organic loading rates (OLR 5–20 kg COD m^{-3}/d). Applying the modified Stover-Kincannon model to the reactor, the maximum removal rate constant (U_{max}) and saturation value constant (KB) was found to be 2 kg/m^3/d and 1.69 kg m^3/d, respectively. Among the advantages of an EGSB in relation to a USB is the ability to work with higher organic loads and the dilution of inhibitors due to recycling. On the other hand, it requires leaner reactors, a more complex system of phase separation, and more power to achieve fluidization. A lab-scale anaerobic hybrid (combining sludge blanket and filter) reactor was operated in a continuous mode to study anaerobic biodegradation of distillery spent wash (Kumar et al. 2007). The study demonstrated that at HRT 5 days and OLR 8.7 kg COD/m^{3}d, the COD removal efficiency of the reactor was 79%. The anaerobic reduction of sulfate increases sulfide concentration, which inhibited the metabolism of methanogens and reduced the performance of the reactors. deBarros et al. (2017) improved methane production from sugarcane vinasse with filter cake in thermophilic UASB reactors, with predominance of *Methanothermobacter* and *Methanosarcina* archaea and *Thermotogae* bacteria. This study demonstrates increased methane production from vinasse through the use of sugarcane filter cake and improved effluent recirculation, with elevated OLR and good reactor stability. We used UASB reactors in a two-stage configuration, with OLRs up to 45 g COD/L d, and obtained methane production as high as 3 L/L/d. Quantitative PCR indicated balanced amounts of bacteria and archaea in the sludge, and of the predominant archaea orders, Methanobacteriales and Methanosarcinales. 16S rDNA sequencing also indicated the thermophilic Thermotogae as the most abundant class of bacteria in the sludge.

7.3.5 Granular-Bed Anaerobic Baffled Reactor

The anaerobic baffled reactor (ABR) presents certain advantages that make it suitable for the anaerobic treatment of diverse types of wastewaters. The ABR is compartmentalized horizontally, using a series of vertical baffles, which force the wastewater to flow over and under the baffles as it travels from the inlet to the outlet, as shown in Figure 7.5. ABR is one of the high rates of anaerobic wastewater treatment systems. It is unique from other high rate anaerobic systems because of its compartmentalized nature, which encourages phase separation along the length of the reactor. The ABR contains a number of compartments that are separated by baffles, thus giving a plug flow system, and minimizes biomass washout (Barber and Stuckey 1999). Each compartment of ABR by itself is believed to be a separate treatment unit and completely mixed. Furthermore, the reactor construction is simple, with no separate gas collection unit. ABR can achieve high volumetric OLRs maintaining at the same time long SRT independent of the HRT. The ABR, first developed by Bachmann (1983), has numerous advantages, including good resilience to hydraulic and organic shock loads, long biomass retention times, low sludge yields, and the ability to separate the various phases of anaerobic catabolism longitudinally down the reactor. The advantages make a shift in bacterial population, and allow increased protection against toxic materials and high resistance to changes in environmental parameters such as pH and temperature. In addition, it has a simple design and cheap construction, appropriate to apply in industry, and it has better COD and oil removal than UASB during the treatment of recalcitrant organic wastewater and constructed containing hydrocarbon wastewater. Due to these potential problems associated with the traditional UASB operation, the use of an ABR coupled with the UASB concept is being proposed for the treatment of high-strength wastewaters, here referred to as "granular-bed baffled reactor". GRABBR is one of the high-rate, gravity-flow systems for anaerobic treatment of wastewaters.

It combines the advantages of the ABR and the UASB by utilizing granular biomass and anaerobic phase separation characteristics. Despite all the advantages of the ABR system, reactor performance in comparison with other anaerobic systems has been found relatively poor at high organic loading rate or short HRT. At short HRTs, biomass losses could pose a serious problem and eventually lead to poor treatment. Short HRT could also encourage short-circuiting that could result in inferior reactor performance. In order to withstand higher loading rates at short HRTs in the baffled system without compromising the effluent quality, it is essential to use biomass that is not only highly active but also structurally stable and possesses good settling characteristics. This will also discourage significant intercompartmental

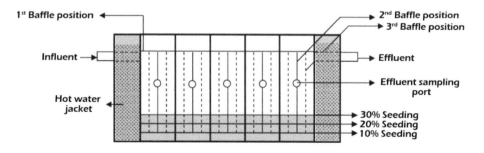

FIGURE 7.5 A schematic view of the granular-bed baffled reactor.

biomass transfer and consequently minimizes biomass losses. Anaerobic granular sludge has been reported to have significantly higher specific methanogenic activity compared to non-granular sludge, which results in high treatment performance. Oosterkamp et al. (2016) showed that thin stillage contains easily degradable compounds suitable for anaerobic digestion and that hybrid reactors can efficiently convert thin stillage to methane under mesophilic and thermophilic conditions. The methane production rate was 0.2 L/g COD, with a methane percentage of 60% and 64%, and 92% and 94% soluble COD removed, respectively, by the mesophilic and thermophilic reactors. Furthermore, we found that optimal conditions for biological treatment of thin stillage were similar for both mesophilic and thermophilic reactors. Bar-coded pyrosequencing of the 16S rRNA gene identified different microbial communities in mesophilic and thermophilic reactors and these differences in the microbial communities could be linked to the composition of the thin stillage.

8

Phytoremediation: An Eco-Sustainable Green Technology for Remediation and Restoration of Distillery Waste Contaminated Environment

8.1 Introduction

For developing countries such as India, China, and Brazil, the ethanol-producing distillery industry plays an indispensable role in benefiting the employment problem, fiscal revenue, and economic development. During the process of ethanol production, a large volume of liquid waste rejects comes out of the distillery industry as a spent wash. This high-strength toxic liquid waste (spent wash) is a discharge from the fermentative production of ethanol. Spent wash is characterized with high levels of recalcitrant pollutants in the form of intense color, low pH, and high chemical oxygen demand (COD), biological oxygen demand (BOD), ash content, inorganic, organic chemicals and dissolved salts (Alves et al. 2015; Chavan et al. 2013; David et al. 2016). The color associated with spent wash not only causes aesthetic damage to water bodies but also prevents the penetration of sunlight through water, which leads to a reduction in the rate of photosynthesis and dissolved oxygen levels, which further creates unwanted anaerobic conditions in the aquatic systems, leading to the death of aquatic biota (Kumar and Gopal 2001; Ramakritinan et al. 2005; Ayyasamy et al. 2008). Distillery effluent also affects agricultural crops in cases where it is directly disposed of into the soil by causing soil acidification (Jain et al. 2005; Kaushik et al. 2005; Jain and Srivastava 2012; Narain, et al. 2012). Although physicochemical methods of effluent treatment are easy to use, these methods may not be always cost-effective and environment-friendly and cannot be employed in all distilleries (Asaithambi et al. 2012; Liakos and Lazaridis, 2014; Prajapati et al. 2016; Tripathy et al. 2020). Aerobic systems employing fungi, bacteria, and algae have been investigated for the removal of melanoidins from distillery effluent (Mohanakrishna et al. 2009; Jack et al. 2014) However, these systems require additional nutrients, which increase the cost of the treatment. On the other hand, there are numerous conventional biological treatment processes such as activated sludge and biomethanation that are cheaper than physicochemical methods such as adsorption (activated carbon and resins), coagulation-flocculation, or advanced oxidation processes, which include ozonation, ultraviolet radiation, and Fenton processes, have also been effective in removing the color from stillage (Mohana et al. 2007; Kumar et al. 1998; Miyata et al. 2000; Ahansazan et al. 2014; David et al. 2016; Kaushik et al. 2018). Moreover, their main disadvantages are the excessive use of chemicals, sludge production with subsequent disposal problems, and high operational costs when considering their application to effluent on a large scale. Also, high sulfate, phosphate, and nitrogen content, as well as colorants (e.g., polyphenols, caramels, and melanoidins) in the distillery effluent, make it difficult to treat (Rani et al. 2013; España-Gamboa et al. 2015; Manyuchi et al. 2018; Kumar et al. 2020a). The melanoidins in distillery effluent possessed antimicrobial activities, which are recalcitrant-coloring pollutants that cause serious environmental problems and health threats in human and animals as well as decreasing biodegradation in the treatment (Chandra et al. 2008; Bharagava et al. 2009; Fan et al. 2011; Arimi et al. 2014; Chandra et al. 2018a). In many instances, industries dilute the anaerobically treated distillery spent wash by mixing with raw water before discharge in order to meet the set waste disposal standards. This dilution, even though accepted in some regions, is of great environmental concern as it does not reduce the absolute pollution load of the effluent. Thus, the development of suitable treatment technologies efficient for the removal of color and reduction of toxicity of distillery effluents is important (Alves et al. 2015; Arimi et al. 2015; Chowdhary et al. 2018a). Besides the generation of anaerobically treated spent wash, distilleries produce a huge quantity of sludge as a solid waste per day (Srivastava et al. 2009; Chandra and Kumar 2017b). Distillery sludge is characterized by high organic matter (OM), phenol, sulphate, inorganic compounds (i.e., Na^+, Cl^-, SO_4^{2-}, PO_3^{4-}), and heavy metals such as iron (Fe), zinc (Zn), copper (Cu), chromium (Cr), cadmium (Cd), manganese (Mn), nickel (Ni), and lead (Pb) present in high quantity, i.e., Fe: 2403.64 mg/kg, Zn: 210.624 mg/kg, Cu: 73.63 mg/kg, Cr: 21.84 mg/kg, Cd: 1.446 mg/kg, Mn: 126.292 mg/kg, Ni:13.425 mg/kg, and Pb: 16.332 mg/kg, along with melanoidins (Chandra and Kumar 2017c). Distillery sludge also contains diverse organic pollutants such as dodecanoic acid, octadecanoic acid, *n*-pentadecanoic acid, hexadecanoic acid, β-sitosterol, stigmasterol, β-sitosterol trimethyl ether, heptacosane, dotriacontane, lanosta-8, 24-dien-3-one, 1-methylene-3-methyl butanol, and 1-phenyl-1-propanol as androgenic and mutagenic compounds, which are potentially harmful to human health and wildlife animals (Chandra and Kumar 2017b; Chandra et al. 2018b). In addition, organic pollutants are toxic to plants and microorganisms; the presence of organic pollutants in soil and water decreases plant growth and its phytoremediation efficacy. The organic pollutants bind with various metallic ions to make an organometallic complex that resulted in enhancing the vulnerability of organometallic complex toward its toxicity in the environment (Kumar and Chandra 2020b). Melanoidins are major nitrogenous polymeric, acidic, negatively charged, and highly dispersed colloids organic compounds present in distillery waste (Ahmed et al. 2020). Melanoidins have also antioxidant properties, which render them toxic to microorganisms and recalcitrance to biological wastewater treatments (Caderby et al. 2013; Arimi et al. 2014; Kaushik et al. 2018). It has been reported that various heavy metal ions such as Cu^{2+}, Cr^{3+}, Fe^{3+}, Zn^{2+}, Mn^{2+}, Co^{2+}, and Pb^{2+} bind with melanoidins to form an organometallic complex (Hatano and Yamatsu, 2018; Kumar and Chandra 2020b). This reflects the magnitude of the environmental pollution caused by the waste generated from the distillery sector all over India, as shown in Figure 8.1.

For the removal of organic maters from distillery effluent and sludge both, the most useful industrial treatment technology is bioremediation because it can be performed on-site, at a lower cost, with limited inconveniences and minimal environmental impact, it eliminates the waste permanently, and it can be used in conjunction with various physical and chemical treatments (Kumar et al. 2018). However, this process requires long times and specific treatment conditions and, in addition, heavy metals, and some biorefractory compounds are not suitable for bioremediation. Moreover, there are still some problems in applying these approaches to raw effluent treatment due to the death of microbial cells in high concentrations of toxic organic and inorganic pollutants. Biological processes are usually aimed at removing readily biodegradable organic compounds or relatively readily hydrolyzable substances but are not specifically designed to remove heavy metals and refractory organic chemicals from complex effluent. Phytoremediation is an ideal approach for the treatment and/or elimination of toxic organic, inorganic, and organometallic pollutants from the contaminated environment or to render them harmless (Chandra and Kumar 2018b; Kumar 2021; Kumar et al. 2021c). It has been considered the most promising technology for distillery waste management due to its minimal site disturbance, low-cost, and higher public acceptance when compared with conventional remediation methods (Chandra et al. 2015; Chandra and Kumar 2015b). Phytoremediation employs plants, alone or together with their associated microorganisms, to clean up and/or improve soil and water quality by inactivation or translocation of pollutants in different parts of the plant, without negative impacts on the structure, fertility, or biological activity of the soil (Kumar and Chandra 2020c; 2018c). Thus, phytoremediation, together with the use of microbes, facilitates greater remediation of heavy metals as compared to phytoremediation alone. Figure 8.2 shows the view of plant-based phytoremediation technology in remediation of industrial-waste-contaminated soil. Generally, plants use varies mechanisms to uptake different organic and inorganic pollutants, which make the basis of phytoremediation technology. They employ numerous kinetic processes, including phytoextraction, phytostabilization, phytovolatilization, phytodegradation, and rhizofiltration, as illustrated in Figure 8.3.

This chapter presents a comprehensive review of role of plant-bacteria synergism in the removal/degradation of pollutants from distillery effluent in terrestrial and aquatic environment.

FIGURE 8.1 Generation of distillery waste and its impact on the environment (a) Spent wash discharges during ethanol production (b) View of anaerobic digestor installed in premises of distillery industry (c) The brown colored spent wash discharges after biomethanation of spent wash. (d and e) The sludge solid disposed as a waste product during biomethanation of spent wash (f–j) A view of aquatic and soil pollution due to indiscriminate disposal of untreated or partly treated distillery spent wash.

8.2 Phytoremediation and Their Different Processes

In view of the widespread contamination of the environment by persistent and toxic chemical pollutants discharged from distilleries, it is essential to develop effective and eco-friendly methods for their remediation. Phytoremediation is a promising, inexpensive, and eco-friendly rehabilitation approach that uses a broad range of plants and their related enzymes, and associated microbes for remediating pollutants through physical, chemical, and biological processes present in different environmental matrices (Hatano and Yamatsu 2018; Chandra et al. 2018b). They can metabolize, detoxify, and/or

FIGURE 8.2 A view of phytoremediation of industrial-waste-polluted sites (a) The luxuriant and healthy growth of green plant in contaminated soil indicates the phytoremediation efficiency of plant (b and c) A large view shows the texture of contaminated soil.

biotransform many refractory pollutants either to obtain carbon and/or energy for their growth or as cosubstrates, thus converting them to simpler products such as carbon dioxide (CO_2) and/or water (H_2O). Phytoremediation technology can be divided into two groups based on the physical location of the remedial action: (i) in situ phytoremediation and (ii) ex situ phytoremediation. In situ phytoremediation involves the placement of green plants in contaminated soil or sediment that is in contact with contaminated groundwater for the purpose of remediation (Hatano et al. 2016; Kumar and Chandra 2020c). In this approach, the contaminated material is not removed prior to phytoremediation. For polluted sites where the pollutants are not bioavailable to plants, such as pollutants present in deep aquifers, an alternative method of remediation applying ex situ phytoremediation is possible. In this approach, the contaminated media are removed from the actual site using mechanical means and then transferred to a temporary treatment area where they can be exposed to plants selected for phytoremediation. After treatment (phytoremediation), the cleaned soil or water can be transferred to the original site, and the plant may be harvested for disposal if necessary. Several native plant species that grow on distillery waste-contaminated sites under natural environment have indicated the phytoremediation potential as natural hyperaccumulators of heavy metals from complex organic wastes because these plants are naturally adapted in terms of growth, survival, and reproduction under the environmental stresses, compared to plants introduced from another environment (Figure 8.2). Plants can accumulate organic pollutants from contaminated sites and detoxify them through their metabolic activities.

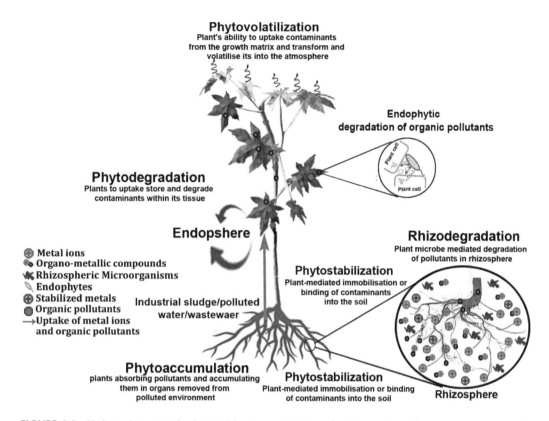

FIGURE 8.3 Various plant strategies involved in phytoremediation of polluted soil and/or water contaminated with organic and inorganic or organometallic compounds (Kumar and Chandra 2018b).

From this point of view, green plants can be regarded as a "green liver" for the biosphere (Sandermann 1994). Plants naturally provide roots, stems, and leaves as habitats for a wide array of microorganisms that simultaneously can breakdown contaminants enhancing the treatment process. Hence, depending upon the detoxification process, applicability, medium, and type and extent of pollution, phytoremediation processes can be subdivided into the following categories.

8.2.1 Phytoextraction

Phytoextraction employs metal-accumulating plants. It can be performed by three means: (i) phytoextraction by trees, (ii) phytoextraction by crops, and (iii) phytoextraction by grasses. Each has its own advantages and disadvantages that take up toxic metals from contaminated media and concentrate them in the harvestable biomass.

8.2.2 Rhizofiltration

Rhizofiltration is a root zone in situ or ex situ technology, used for the elimination of heavy metals from water and aqueous waste streams that are retained only within the roots of aquatic plants. It reduces the mobility of heavy metals and prevents their migration to the groundwater, thus reducing bioavailability for entry into the food chain.

8.2.3 Phytodegradation

Phytodegradation is a process of absorption and uptake of pollutants by the root system, resulting in metabolic or enzymatic transformation within or external to plants.

8.2.4 Rhizodegradation

Rhizodegradation, also called phytostimulation, involves secretion of root exudates or enzymes in the rhizosphere (root zone) and subsequent microbial degradation of contaminants.

8.2.5 Phytostabilization

Phytostabilization involves the use of certain plants to reduce the mobility and bioavailability of pollutants in the environment, thus preventing their migration to groundwater or their entry into the food chain.

8.2.6 Phytovolatilization

Phytovolatilization involves the plant uptake of metals from soil and their volatilization from the foliage. These plants can subsequently be harvested, processed, or disposed of safely. It is a specialized form of phytotransformation that can be used only for those pollutants that are highly volatile like mercury (Hg) and selenium (Se). Figure 8.3 outlines the common processes involved in phytoremediation of contaminated environment.

8.3　Success Stories on Phytoremediation of Distillery Waste

Phytoremediation can be understood as the ability of plants to degrade, extract, transform, and detoxify through their enzymes and associated microflora, the contaminants of the air, soil, sediments, surface water, and groundwater. It is a green technology that uses plant systems for remediation and rehabilitation of the contaminated sites. Plants have inbuilt enzymatic machinery capable of accumulating and degrading complex structural pollutants and can be used for cleaning of the distillery waste contaminated sites. Figure 8.3 provides a view of plants that are able to promote phytoremediation of distillery waste employing the principles of rhizofiltration, phytoextraction (phytoaccumulation), phytodegradation, and rhizodegradation. Singh et al. (2005) have assessed the ability of aquatic macrophyte *Potamogenton pectinatus* L. to accumulate Fe, Cu, Zn, and Mn from distillery effluent and related toxicity therein. They showed that *P. pectinatus* L. accumulate a significant amount of Cu, Fe, Zn, and Mn from distillery effluent in their tissues during two weeks. Olguín et al. (2007) assessed the phytoremediation potential of *Salvinia minima* baker compared to *Spirodela polyrrhiza* in high-strength synthetic organic wastewater and also evaluated the growth characteristics of *S. minima* in various culture media, including anaerobic effluents from pig wastewater. The authors concluded that *S. minima* are a better option than *S. polyrrhiza* for treating high-strength organic wastewater. Sharma et al. (2011) screened 16 plant species belonging to glycophytes (6 species of *Acacia, Dendrocalamus strictus*, and *Lawsonia inermis*), halophytes (*Atriplex nummularia, Chenopodium murale*, and *Suaeda nudiflora*), and helophytes (*Arundo donax, Phragmites karka, Typha angustata, Scirpus tuberosus*, and *Scirpus littoralis*) to identify their tolerance and capacity at field phytoremediation of biomethanated diluted spent wash in the high-rate evapotranspiration system (HRES). They observed tolerant of *Acacia farnesiana, D. strictus*, and *L. inermis* (glycophytes), *A. nummularia* and *S. nudiflora* (halophytes), and *A. donax* and *P. karka* (helophytes) to the diluted spent wash. Besides pollution abatement and carbon locking, accrued returns of planting these species in HRES are fodder, fuel wood, and biomass for multiple uses and products such as henna and oil (*Atriplex* seeds). Bharagava et al. (2008) studied the metal accumulation efficiency and its physiological effects in *Brassica nigra* L. plants grown in soil irrigated with different concentrations (25%, 50%, 75%, 100% v/v) of post-methanated distillery effluent (PMDE) after 30, 60, and 90 days treatment. *B. nigra* L. accumulate elevated concentration of Zn, Ni, Mn, Fe, Cu, and Cd in their above- and below-ground parts due to increased amount of cysteine and ascorbic acid (work as antioxidants) in leaves, shoot, and root of *B. nigra* L. at all the concentration and exposure periods of PMDE except at 90 days period, where a decrease was observed at 100% PMDE concentration as compared to their respective control. A two-step sequential treatment for sugarcane molasses-based anaerobically treated distillery effluent was reported by Pant and Adholeya (2009a). In the first step, distillery effluent was treated in a hydroponic-based system using two plant species, namely, *Vetiveria zizanioides* and *P. karka* to reduce (up to 84%) the high nitrogen content of the effluent. This first-step hydroponically treated distillery effluent was subjected for treatment by two fungus species in a bioreactor. Decolorization of effluent up to 86.33% was obtained with *Pleurotus florida* Eger EM1303 followed by *Aspergillus flavus* TERIDB9 (74.67%) with a significant reduction in COD as well. This study recommended the distillery effluent treatment without the need of high dilutions and the addition of supplementary carbon sources. Restoration of habitats and in situ cleanup of contaminants can be achieved with significantly reduced remedial costs by this phytotechnology. The heavy metals accumulation potential of *Typha angustifolia* and *Cyperus esculentus* growing in distillery and tannery effluent-polluted wetland sites have been reported by Yadav and Chandra (2011). The metal accumulation pattern in both macrophytes indicated that both macrophytes were noted as a root accumulators for Fe, Cr, Pb, Cu, and Cd. Simultaneously, chlorophyll, protein, cysteine, and ascorbic acid were also induced in *T. angustifolia* than *C. esculentus*. In addition, anatomical observation through transmission electron microscopy in the root of *T. angustifolia* did not show any remarkable changes even after a higher accumulation of various metals in the roots. However, the formation of multinucleolus in a shoot of *T. angustifolia* was found as evidence of extra protein synthesis for tolerance under stress conditions. Hence, *C. esculentus* was observed as potential but less tolerant for metals than *T. angustifolia*. The various plant species reported in phytoremediation of distillery waste polluted sites are presented in Figure 8.4.

Hatano and Yamatsu (2018) evaluated the facilitatory influence of melanoidin-like product (MLP) on phytoextraction efficiency in a medium, including Cd^{2+} or Pb^{2+}, the concentrations of which were adjusted near the regulation values of the Act in Japan. In this study, three *Brassica* species were tested due to their fast growth, high biomass productivity, and high heavy metal absorption, and the cultivation period was two months under sunlight. It was observed that both biomass and Pb uptake were significantly increased by the addition of MLP, and almost all of the Pb was accumulated in the root tissue. Therefore, MLP was able both to detoxify lead ions and to improve their bioavailability in *Brassica* species. A research group from Japan has demonstrated that MLP from sugarcane molasses possess the potential for an accelerator of phytoextraction efficiency of Japanese radish in the Cu-contaminated media (Hatano et al. 2016). It is stated that MLP binds to all the metal ions examined and the binding capacity of MLP toward Cu^{2+} seems to be the highest among them. The metal detoxification by MLP followed the order of $Pb^{2+} > Zn^{2+} > Ni^{2+} > Cu^{2+} > Fe^{2+} > Cd^{2+} > Co^{2+}$. Chowdhary et al. (2018b) investigated the effects of potentially toxic elements on biochemical parameters in *Triticum aestivum* L. and *Brassica juncea* L. plants growing at distillery and tannery wastewater contaminated sites. The analysis of plants showed the highest accumulation of Fe (361 mg/kg in wheat root and 359 mg/kg in mustard leaves) followed by Zn, Cr, and Mn in leaf > shoot > root. Further, the Chl-a, Chl-b, and carotenoids content was also found high in plant samples. Photosynthetic content in wheat and mustard growing at tannery wastewater contaminated sites was Chl-a 3.92, 4.53 (mg/g fw), Chl-b 2.39, 1.29 (mg/g fw), and carotenoids 0.28, 0.32 (mg/g fw). Whereas photosynthetic content in these plants with distillery waste was Chl-a 3.43, 4.88 (mg/g fw), Chl-b 1.12, 2.05 (mg/g fw), and carotenoids 0.24, 0.29 (mg/g fw). In addition, the activity of plant enzymes such as SOD, APx, GPX, MDA, H_2O_2, and catalase was also higher in selected plants in comparison to control plants. A group of researchers from Mexico reported on the use of phytofiltration technology based on *Azolla* sp., a free-floating and fast-growing aquatic plant, for treatment of anaerobically digested sugarcane ethanol stillage (Sanchez-Galvan and Bolanos-Santiago 2018). They demonstrated that developed phytofiltration is efficient for reducing the OM content, nutrients, and color intensity significantly in anaerobically digested stillage (ADS) under the tested conditions. The authors suggested that the conversion of nutrients from ADS into valuable *Azolla* biomass may provide an effective way to produce a very attractive feedstock for the production of a wide spectrum of biofuels. Single-phase system has provided good organic removal efficiency; however, a biphasic system is capable of optimizing the fermentation steps of each stage in separate fermenters. Wetlands are an ecologically friendly way of removing residual Cu from distillery wastewater, as described in the case study by Murphy et al. (2009). These wetlands generally include a wide range of native wetland plants and willow trees,

FIGURE 8.4 The luxuriant growth of native plants on dumped distillery waste indicated the phytoremediation efficiency of plants (a and b) Anaerobically digested distillery sludge dumped in open environment (c) *Phragmites australis* Ld *Solanum nigrum* (e) *Typha latifolia* (f) *Dhatura stramonium* (g) *Parthenium hysterophorous.* (h) *Basella alba* (i) *Cannabis sativa.*

which bind Cu onto their roots, rhizomes, or woody material, thus preventing its release into the environment. A laboratory-scale CW employing *Typha latifolia* was used to treat diluted distillery effluent (Trivedy and Nakate 2000). A root zone of $1.5 \times 0.3 \times 0.3$ m, filled with 75% sand and gravel and 25% soil, was used and the diluted effluent was applied after 4 weeks of planting. The system resulted in 76% COD reduction in 7 days which increased marginally to 78% COD reduction in 10 days. The BOD reduction was 22% and 47% on day 7 and 10, respectively. In yet another instance, a distillery in northern India is presently employing CWs for polishing the effluent prior to land discharge for irrigation in the surrounding paddy fields. The effluent is initially subjected to primary treatment that includes settling and anaerobic digestion in a structured media attached growth type anaerobic reactor. The primary treated effluent, with a COD of 28,000–35,000 mg/L, is subjected to two-stage aeration to bring down the COD to 400 mg/L. Thereafter, it is directed to a CW before the final discharge. Billore et al. (2001) have proved that a four-celled horizontal subsurface flow (HSF) CWs in a multi-cell system are a cost-effective alternative for the treatment of stillage from sugarcane. In this system, plants (*T. latifolia* and *P. karka*) received diluted secondary effluents previously subjected to anaerobic digestion and aeration. The post-anaerobic treated effluent had BOD of 2500 mg/L and COD 14,000 mg/L. A pretreatment chamber filled with gravel was used to capture the suspended solids. All the cells were filled with gravel up to varying heights and cells, third and fourth were planted with *T. latifolia* and *P. karka*, respectively. The overall retention time was 14.4 days and the treatment resulted in 64%, 85%, 42%, and 79% reduction in COD, BOD, total solids, and phosphorus, respectively. This study concluded that CW is a sustainable tertiary treatment technique for the remediation of contaminants present in distillery effluent. In another study, Trivedy and Nakate (2000) used wetland plant *T. latifolia* for treatment of distillery effluent in a CW treatment system. This treatment system resulted in 47% and 78% decrease in BOD and COD, respectively, in an incubation of 10 days. Increasing concentration of distillery effluent significantly reduced the biomass of growing plant with the highest accumulation of Fe being recorded in plants growing in 100% concentration of effluent. Aquatic macrophyte *Potamogeton pectinatus* was used to accumulate Mn, Zn, Cu, and Fe and efficiently clear out the distillery effluent (Singh et al. 2005). Similarly, Olguìn et al. (2008) evaluated the performance of subsurface flow constructed wetlands (SSF-CWs) mesocosms planted with *Pontederia sagittata* and operating at two hydraulic retention times (HRTs), compared to an unplanted SSF-CWs, for the treatment of diluted stillage subjected to no pretreatment apart from an adjustment to pH 6. The planted SSF-CWs were able to remove COD in the range of 80.24–80.62%, BOD_5 in the range of 82.20–87.31%, total Kjeldahl nitrogen (TKN) in the range of 73.42–76.07%, nitrates 56–58.74%, and sulfates 68.58–69.45%, depending on the HRT. However, phosphate and potassium were not removed. A benchscale study conducted for polishing of biomethanated spent wash (primary treated) in constructed wetland planted with *P. karka* (Singh et al., 2010). The reduction in COD (54–63%) and BOD (58–70%) values and their loads (COD: 78–98%, BOD: 81–95%) in the effluents were significant and independent of influent concentrations, whereas color removal (34–82%) was concentration dependent. Mulidzi (2010) operated a CW in terms of COD and other elements removal from winery and distillery wastewater. Results also showed reasonable removal of other elements: potassium, pH, nitrogen, electrical conductivity, calcium, sodium, magnesium, and boron from the wastewater by CWs. Mulidzi (2010) showed the impact of shorter retention time on the performance of CWs in terms of BOD, COD, and other elements removal. The results had shown an overall 60% COD removal throughout the year. This study also indicated the significant removal of other elements: potassium, nitrogen, electrical conductivity, calcium, sodium, magnesium, and boron from distillery wastewater by CWs. The success of CWs for treating toxic and high-strength organic matter wastewater depends on various factors, mainly those related to the organic load, hydraulic retention time, tolerance of the selected plant to the toxic components, and biofilm performance. Kumar and Chandra (2004) showed that a previous detoxification of anaerobic stillage effluents with *Bacillus thuringiensis* enhanced the phytoremediation potential of *Spirodela polyrrhiza*, working in microcosms. Kumar and Chandra (2004) successfully treated anaerobically digested distillery effluent in a two-stage treatment process involving the transformation of recalcitrant-coloring components of the effluent by aerobic bacterium *B. thuringiensis* followed by a subsequent decline of a remaining load of pollutants by a macrophyte *S. polyrrhiza* Schleiden. The biphasic system is the most appropriate treatment method for high-strength wastewater. A combination of bacterial pretreatment followed by CWs plant treatment system was investigated to determine its effect on the removal of heavy metals and detoxification of post-methanated distillery effluent (PMDE). This biphasic treatment of the effluent with *T. angustata* and *B. thuringiensis* removed large quantities of various heavy metals at a range of effluent concentrations (i.e., 10%, 30%, 50%, and 100%). A similar biphasic (two-step) treatment of the PMDE was carried out in a CW with *B. thuringiensis* followed by *T. angustata* L. by Chandra et al. (2008), which resulted in 98–99% COD, BOD, and color reduction after 7 days. The authors recommended that the bacterial pretreatment of PMDE integrated with CW will improve the treatment process of PMDE and promote safe disposal of hazardous distillery waste. Sa et al. (2008) treated diluted sugarcane molasses stillage in CW mesocosms planted with *Pontederia sagittate*. The planted CWs were able to remove COD in the range of 80.24–80.62%, BOD_5 in the range of 82.20–87.31%, TKN in the range of 73.42–76.07%, nitrates in the range of 56–58.74%, and sulfates in the range of 68.58–69.45%, depending on the HRT. Phosphate and potassium were not removed. Chandra et al. (2012) reported 96% and 94.5% reduction in COD and BOD values in a two-step sequential treatment of PMDE by bacteria and wetland plant *Phragmites communis*. He also characterized rhizosphere bacterial communities of *P. communis* and metabolic products generated during the sequential treatment of PMDE in CWs plant treatment system. It is important to use the native plants of contaminated sites for phytoremediation because these plants are naturally adapted in terms of survival, growth, and reproduction under the environmental stresses compared with plants introduced from another environment. Several potential indigenous plants that grow on distillery waste-contaminated sites under natural conditions have indicated the phytoremediation potential of the use of such plants at contaminated sites as natural hyperaccumulators of heavy metals from complex organic wastes (Figure 8.4). Several authors recommended that native plants are a good remediator used for in situ phytoextraction of heavy metals from industrial-waste-polluted sites. Chandra and Yadav (2010) conducted a pot culture experiment to evaluate the accumulation pattern of Cu, Pb, Ni, Fe, Mn, and Zn in *T. angustifolia* grown in Zn-, Mn-, Fe-, Ni-, Pb-, and Cu-rich aqueous solutions of phenols and melanoidins. They concluded that *T. angustifolia* could be an efficient phytoremediator for heavy metals from melanoidin-, phenol-, and metal-containing industrial effluent at optimized conditions. Yadav and Chandra (2011) analyzed the heavy metals accumulation and ecophysiological effect on *T. angustifolia* L. and *Cyperus esculentus* L. growing in distillery and tannery effluent-polluted natural wetland site, Unnao, India. Both macrophytes found potential Fe, Cr, Pb, Cu, and Cd in their root and shoots. Simultaneously, chlorophyll, protein, cysteine, and ascorbic acid were also induced in *T. angustifolia* than *C. esculentus*. In addition, formation of multinucleolus in shoot of *T. angustifolia* was found as evidence of extra protein synthesis for tolerance under stress conditions. Hence, *C. esculentus* was observed having potential but less tolerance for metals than *T. angustifolia*. The study concluded that *T. angustifolia* had a higher potential for heavy metals accumulation than *C. esculentus* from distillery and tannery wastewater. Hence, these two plants could be used for the phytoremediation of the heavy-metals-contaminated swampy lands and wastewater. Phytoextraction of heavy metals is performed by potential native plants and their microscopic observation of root growing on stabilized distillery sludge as a prospective tool for in situ phytoremediation of industrial waste (Chandra & Kumar, 2017c). The study has revealed that PMDS contained high amounts of Fe, Zn, Mn, Cu, Cr, Pb, and Ni along with melanoidins and other co-pollutants. The phytoextraction pattern in 15 potential native plants growing on sludge showed that the *Blumea lacera, Parthenium hysterophorous, Setaria viridis, Chenopodium album, Cannabis sativa, Basella alba, Tricosanthes dioica, Amaranthus spinosus* L., *Achyranthes* sp., *Dhatura stramonium, Sacchrum munja*, and *Croton bonplandianum* accumulated a significant amount of heavy metals above their threshold limits. The BCF of all native plants were found <1, while TF was noted in most of the growing plants >1. This gives strong evidence for the phytoextraction and in situ remediation potential of these native plants for these heavy metals from the contaminated site. This gives a strong evidence of hyperaccumulation for the tested metals from complex distillery waste. Furthermore, the TEM observations of root of *P. hysterophorous, C. sativa, S. nigrum*, and *R. communis* showed the formation of multi-nucleolus, multi-vacuoles, and deposition of metal granules in cellular component of roots as a plant adaptation mechanism for phytoextraction of heavy metal-rich polluted site. Hence, these native plants may be used as a tool for in situ phytoremediation and eco-restoration of industrial-waste-contaminated site. The TEM observation in root of representative native plants revealed that tested hyperaccumulator plants showed the formation of the multi-nucleolus, multi-vacuoles, and deposition of metal granules in their cellular components as a process of adaptation at heavy-metal-containing organic waste-rich polluted site in the environment. Thus, the study recommended that these plants can be used for minimization of heavy metals from contaminated soil, but it should be prohibited for use as food and fodder due to health hazards; various heavy metals may be recovered from specific parts of plants through phytomining. Chandra et al. (2018b) evaluated the assessed heavy metal phytoextraction potential of native weeds and grasses from endocrine-disrupting chemicals-rich complex distillery sludge and their histological observations during in situ phytoremediation. Authors collected nine native plant species (grasses and weeds)—*B. alba, Calotropis procera, Tinospora cordifolia, Rumex dentatus, C. album, Pennisetum purpureum, Cynodon dactylon, S. munja*, and *Argemone mexicana*—which were abundantly grown on disposed distillery sludge containing the mixture of complex organic pollutants benzoic acid, 3,4,5-tris (TMS oxy), TMS ester; stigmasterol TMS ether; hexanedioic acid, dioctyl ester; benzene, 1-ethyl-2-methyl; 5α-cholestane, 4-methylene; campesterol TMS; and β-sitosterol and lanosterol, but also retains a high quantity of Pb, Ni, Fe, Mn, Cu, and Zn that enhances the toxicity of sludge to the environment (Chandra et al. (2018b). This study revealed that distillery sludge contains not only mixture of complex organic pollutants but also retains a high quantity of Fe (5264.49 mg/kg), Zn (43.47 mg/kg), Cu (847.46 mg/kg), Mn (238.47 mg/kg), Ni (15.60 mg/kg), and Pb (31.22 mg/kg), which enhances the toxicity of sludge to the environment. The metal accumulation pattern of weeds and grasses growing naturally in the contaminated environment revealed that all the plant species accumulated a high amount of metals in root, shoot, and leaves. Furthermore, the BCF and TF were found >1 for the majority of plants for various metals. Thus, this gives strong evidence for hyperaccumulation tendency of these native weeds and grasses from the organometallic polluted site mixed with androgenic and mutagenic compounds. The TEM observation in root tissues of all plant species showed apparent formation of multi-nucleus, multi-nucleolus, multi-vacuoles, mitochondria, and dense deposition of metal granules in the cellular organelle of a plant that supported the plant tolerance mechanism at higher concentration of heavy metals in presence of other complex organic pollutants. Thus, the study recommended that these plants can be used for in situ monitoring and phytoextraction of heavy metals from organometallic waste-contaminated sites. Further, it is also recommended that these plants should be prohibited for the use of food and fodder due to health hazards. Hatano et al. (2016) examined the chelating property of MLP to evaluate the facilitatory influence on the phytoextraction efficiency of Japanese radish (*Raphanus sativus* var. longipinnatus) seeds. MLP binds to all the metal ions examined and the binding capacity of MLP toward Cu^{2+} seems to be the highest among them. The metal detoxification by MLP followed the order of $Pb^{2+} > Zn^{2+} > Ni^{2+} > Cu^{2+} > Fe^{2+} > Cd^{2+} > Co^{2+}$. Furthermore, in the phytoextraction experiment using copper sulfate, the application of MLP accelerated the detoxification of copper and the bioavailability in radish sprouts. MLP enhances phytoextraction of lead through three Canola appeared *Brassica* species. MLP were able to both detoxify Pb^{2+} and improve their bioavailability in *Brassica* species. In contrast, only these species with MLP or citric acid survived in the nutrient medium with 1 mM cadmium sulfate. The phytoextraction of Cd^{2+} using these species was therefore impractical under the Act. Finally, the treatment with MLP increased both biomass and Pb uptake, indicating that MLP is ideal as a phytoextraction accelerator for Pb (Hatano and Yamatsu, 2018). They concluded that MLP was able to detoxify Pb^{2+} and to improve their bioavailability in *Brassica* species. Chaturvedi et al. (2006) reported the phytoremediation potential of *P. australis* grown on distillery effluent-contaminated site. She also characterized the diverse bacterial species from the rhizospheric zone of *P. australis*. The culturable bacterial species were helpful for the degradation and detoxification of noxious pollutants that exist in the distillery effluent. They observed 75.5% reduction of color by the same bacterial species along with a concomitant reduction in BOD, COD, sulfate, phenol, and heavy metals values. A recent study conducted by Chandra and Kumar (2017b) using restriction fragment length polymorphism (RFLP) approach explored the microbial communities composition and function during in situ bioremediation of distillery sludge. The results indicated that *Bacillus* sp. and *Enterococcus* were found dominantly growing autochthonous bacterial communities during in situ bioremediation of distillery sludge due to the availability of diverse habitats and metabolization capabilities. Hence, this ability creates a specific niche of these grown bacterial communities that may lead to in situ phytoremediation of organic and inorganic compounds. Several plants growing in polluted soil and water host different types of rhizosphere bacteria able to degrade organic pollutants and have been isolated, and degradation pathways and genes involved in organic pollutants degradation have been identified. Even though these bacteria showed high potential to degrade different persistent organic pollutants (POPs), these are unable to survive and proliferate in the contaminated soils. the luxuriant growth of native plants of stabilized distillery sludge indicates the capabilities of their rhizospheric bacterial communities, which might be detoxifying the complex organic compounds of distillery sludge during in situ phytoremediation (Figure 8.5).

FIGURE 8.5   Distillery sludge dumping site and phytoremediating plant (a) Fresh distillery sludge discharged after anaerobic digestion of spent wash (b) *Saccharum arundinaceum* plant grown on distillery sludge dumping site.

Kumar and Chandra (2020b) explored the rhizospheric bacterial communities structure and composition associated with *Saccharum arundinaceum* grown on organometallic pollutants-rich hazardous distillery sludge. The luxuriant growth of *S. arundinaceum* showed the potential of plant for phytoremediation of disposed of distillery sludge that gives an evidence of mineralization of organometallic complex pollutants (Figure 8.5). Further, metagenomic analysis of rhizospheric sludge of *S. arundinaceum* revealed the presence of *Rheinheimera* (21%), *Sphingobacterium* (17%), *Idiomarina* (8%), *Acidothermus* (4%), uncultured *Bacillus* (4%), *Bacillus* (3%), *Pseudomonas* (2%), *Flavobacterium* (2%), uncultured bacterium (2%), *Parapedobacter* (2%), *Alcanivorax* (2%), *Acholeplasma* (2%), *Hyphomonas* (1%), and *Aquamicrobium* (1%) as dominant taxa as shown in Figure 8.6. The taxonomic information confirmed that the diversity of rhizospheric bacterial communities of *S. arundinaceum* is quite different from the non-rhizospheric sludge communities. These bacteria are likely to contribute

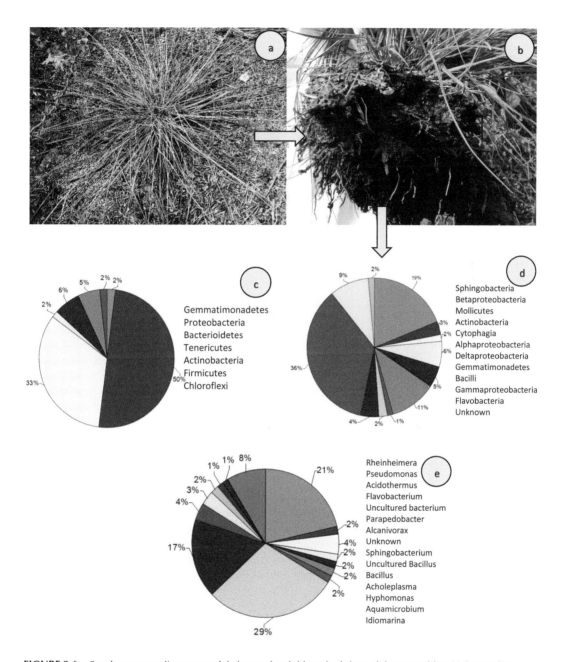

FIGURE 8.6 *Saccharum arundinaceum* and their associated rhizospheric bacterial communities. (a) *S. arundinaceum* grown on distillery sludge dumped site. (b) Roots of *S. arundinaceum* show luxuriant growth. (c–e) Metagenomics analysis of rhizospheric bacterial communities of *S. arundinaceum* grown on organometallic sludge of sugarcane-molasses-based distillery pie chart showing the bacterial communities composition and relative abundance distribution at the different taxa: (a) phylum, (b) class, and (c) genus level that were identified by metagenomics sequencing technique. Percent relative abundance refers to cumulative abundance (Kumar and Chandra 2020b).

to the biodegradation and/or biotransformation of organic pollutants exhibiting in distillery sludge. The comparative GC-MS analysis also supported that the majority of organic pollutants detected in non-rhizospheric sludge were eliminated from rhizospheric sludge samples and few compounds were generated as metabolic products by the combined action of plant and bacterial communities.

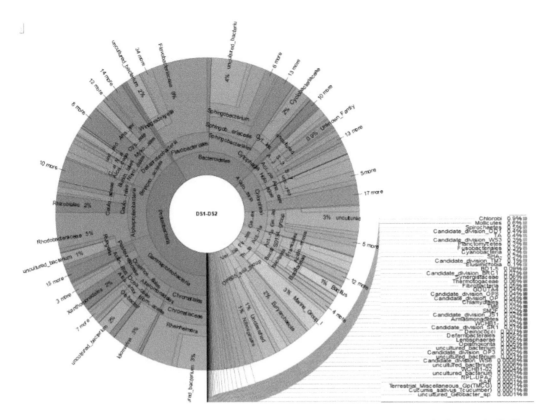

FIGURE 8.7   Krona chart analysis revealed an apparent divergence in bacterial diversity of rhizospheric distillery sludge.

Furthermore, metagenomic visualization in a Krona diagram indices an apparent divergence in bacterial diversity of rhizospheric and non-rhizospheric distillery sludge sample, as illustrated in Figure 8.7. Krona is taxonomy web visualization; circles from inside to outside stand for different classification levels, and the area of the sector means the respective proportion of different OTUs annotation results. **Table 8.1** shows the details of numerous successful studies showing phytoremediation of distillery waste pollutants/or polluted sites using in situ or ex situ experimental approach.

TABLE 8.1

Phytoremediation of Distillery Waste as Reported by Various Researchers Worldwide

Country	Tested Plant Species	Core Findings of the Study	Reference
India	*Typha latifolia*	The diluted effluent applied in CWs vegetated with *T. latifolia* has resulted in 76% COD reduction at day 7, which increased marginally to 78% COD reduction at day 10. The BOD reduction was 22% and 47% on day 7 and 10, respectively	Trivedy and Nakate (2000)
India	*Typha latifolia* and *Phragmites karka*	The post-anaerobic treated effluent had BOD of 2500 mg/L and COD 14,000 mg/L. The treatment resulted in 64%, 85%, 42%, and 79% reduction in COD, BOD, TS, and phosphorus, respectively	Billore et al. (2001)

TABLE 8.1 *(Continued)*

Country	Tested Plant Species	Core Findings of the Study	Reference
India	*Spirodela polyrrhiza* Schleiden	PMDE in a two-stage treatment process involving the transformation of recalcitrant-coloring components of the effluent by aerobic bacterium *Bacillus thuringiensis* followed by a subsequent decline of a remaining load of pollutants by a macrophyte *S. polyrrhiza* Schleiden	Kumar and Chandra (2004)
India	*Potamogenton pectinatus* L.	*Potamogenton pectinatus* L. to accumulate significant amounts of various heavy metals, namely, Fe, Cu, Zn, and Mn from distillery effluent in their tissues during a period of two weeks	Singh et al. (2005)
India	*Phragmites australis*	Isolated and characterized by 15 bacterial strains root-associated rhizospheric bacterial communities of *P. australis* grown on distillery waste-contaminated site. All the isolates grew on effluent-supplemented medium resulting in the reduction of the levels of distillery pollutants and their color by 75.5% Subsequently, there was a reduction in BOD, COD, phenol, sulfate, and HMs	Chaturvedi et al. (2006)
India	*Typha angustata* L.	The biphasic treatment of effluent with *T. angustata* and *B. thuringiensis* (MTCC 4714) removed large quantities of various HMs at a range of effluent concentration (i.e., 10%, 30%, 50%, and 100%)	Chandra et al. (2008)
Mexico	*Pontederia sagittate*	A SSF-CWs mesocosms planted with *P. sagittate* were able to remove COD (80.24–80.62%), BOD_5 (82.20–87.31%), TKN (73.42–76.07%), and nitrates (56–58.74%) and sulfates (68.58–69.45%) from wastewater. However, phosphate and potassium were not removed	Olguin et al. (2008)
India	*Brassica nigra* L.	*B. nigra* L. accumulate elevated concentration of Zn, Ni, Mn, Fe, Cu, and Cd due to increased amount of cysteine and ascorbic acid in their leaves, shoot, and root at all the concentration and exposure periods of PMDE except at 90-days period	Bharagava et al. (2008)
India	*Vetiveria zizanioides* and *Phragmites karka*	A two-step sequential treatment for sugarcane molasses-based anaerobically treated distillery effluent was found effective for color removal. In the first step, distillery effluent was treated in a hydroponic-based system using *V. zizanioides* and *P. karka* to reduce (up to 84%) the high nitrogen content of the effluent. Decolorization of effluent up to 86.33% was obtained with *Pleurotus florida* Eger EM1303 followed by *Aspergillus flavus* TERIDB9 (74.67%) with a significant reduction in COD as well	Pant and Adholeya (2009)
South Africa	*Typha*	The significant removal of other elements, namely, potassium, nitrogen, electrical conductivity, calcium, sodium, magnesium, and boron, from distiller wastewater by CWs as a secondary treatment system was reported	Mulidzi (2010)

(Continued)

TABLE 8.1 *(Continued)*

Country	Tested Plant Species	Core Findings of the Study	Reference
India	*Acacia, Dendrocalamus strictus, Lawsonia inermis, Atriplex nummularia, Chenopodium murale, Suaeda nudiflora Arundo donax, Phragmites karka, Typha angustata, Scirpus tuberosus,* and *Scirpus littoralis*	Out of 16 plant species, *A. farnesiana, D. strictus, L. inermis, A. nummularia, S. nudiflora, A. donax,* and *P. karka* were showed having tolerance to the diluted spent wash in the high-rate evapotranspiration system	Sharma et al. (2011)
India	*Typha angustifolia* and *Cyperus esculentus*	The tested macrophytes were root accumulator for Fe, Cr, Pb, Cu, and Cd. Simultaneously, chlorophyll, protein, cysteine, and ascorbic acid were also induced in *Typha angustifolia* than *C. esculentus.* In addition, anatomical observation through TEM in the root of *T. angustifolia* did not show any remarkable changes even after a higher accumulation of various metals in the roots	Yadav and Chandra (2011)
India	*Phragmites communis*	Authors were reported 96.0% and 94.5% reduction in COD and BOD values in a two-step sequential treatment of PMDE by bacteria and wetland plant *Phragmites communis*	Chandra et al. (2012)
Japan	Japanese radish	MLP binds to all the metal ions examined and the binding capacity of MLP toward Cu^{2+} seems to be the highest among them. The metal detoxification by MLP followed the order of $Pb^{2+} > Zn^{2+} > Ni^{2+} > Cu^{2+} > Fe^{2+} > Cd^{2+} > Co^{2+}$	Hatano et al. (2016)
India	*Dhatura stramonium, Achyranthes* sp., *Kalanchoe pinnata, Trichosanthes dioica, Parthenium hysterophorous, Cannabis sativa, Amaranthus spinosus* L., *Croton bonplandianum, Solanum nigrum, Ricinus communis, Sacchrum munja, Basella alba, Setaria viridis, Chenopodium album,* and *Blumea lacera*	All the tested plants accumulate significant concentration of heavy metals in their roots, shoots, and leaves. TEM observations of the root of *R. communis, Solanum nigrum, C. sativa,* and *P. hysterophorus* showed the formation of multi-vacuoles and multi-nucleolus and deposition of HM granules in cellular component of roots as a plant adaptation mechanism for phytoextraction of HM-rich polluted site	Chandra and Kumar (2017c)
Mexico	*Azolla* sp.	Phytofiltration system involving *Azolla* sp. is efficient for reducing the organic matter content, nutrients, and color intensity significantly in ADS under the tested conditions	Sanchez-Galvan and Bolanos-Santiago (2018)
Japan	*Brassica* sp.	Both biomass and Pb uptake were significantly increased by the addition of MLP, and almost all of the Pb was accumulated in the root tissue. Authors concluded that MLP was able to both detoxify lead ions and improve their bioavailability in *Brassica* species	Hatano and Yamatsu (2018)

TABLE 8.1 *(Continued)*

Country	Tested Plant Species	Core Findings of the Study	Reference
India	*Saccharum munja, Cynodon dactylon, Pennisetum purpureum, Argemone mexicana, Chenopodium album, Rumex dentatus, Tinospora cordifolia, Calotropis procera, and Basella alba*	All the tested plants have been hyperaccumulator for Pb, Ni, Mn, Cu, Zn, and Fe, extracted from organometallic polluted sites mixed with androgenic and mutagenic compounds. Moreover, the root tissues of all plant species observed through TEM analysis showed the formation of multi-vacuoles, multi-nucleolus, multi-nucleus, opaque deposition of HMs granules in the cellular organelle of the plant cell, which indicated the HMs tolerance mechanism of the plant at an elevated concentration of HMs and other complex co-pollutants	Chandra et al. (2018a)

TEM: transmission electron microscopy; MLP: melanoidin-like products; Pb: lead; Ni: nickel; Mn: manganese; Cu: copper; Zn: zinc; Cd: cadmium; Co: cobalt; Fe: iron; HMs: heavy metals; PMDE: post-methanated distillery effluent; SSF-CWs: subsurface flow constructed wetlands; BOD: biological oxygen demand; COD: chemical oxygen demand; TS: total solids; HSWW: high-strength synthetic wastewater; ADS: anaerobically digested stillage.

9

Recycling and Reuse of Distillery Waste by Vermitechnology

9.1 Introduction

Distillery, an important subunit of the sugar production industry, produces massive quantities of complex and variable solid and liquid wastes containing various recalcitrant and toxic chemicals, depending on the processed and the type of raw material used in ethanol production (Srivastava et al. 2009; David and Arivazhagan 2015; David et al. 2016). The majority of molasses-based distilleries generate ~12–15 L of effluent, commonly known as spent wash, per liter of ethanol production. In addition to spent wash generation, a huge amount of solid waste as a yeast sludge is formed in the distilleries, which cause environmental pollution when it is disposed into the environment without adequate treatment. In environmental terms, distillery effluent has a strong color, high pH, conductivity, toxicity, and low biodegradable nature due to high concentrations of complex organic compounds, such as melanoidins, phenolics, androgenic-mutagenic compounds, heavy metals, and other substances as well as inorganic salts content (Jain and Srivastava 2012; Kaushik and Thakur 2013; Chandra and Kumar 2017a; Hoarau et al. 2018). Thus, untreated distillery spent wash can be considered one of the major pollutants for the aquatic and terrestrial ecosystem. Estimates show that organic waste produced from the distillery industry includes effluent of molasses 350,000 L/day; yeast sludge 20,000 L/day, and spent malt grain wash 120,000 L/day. In order to attend the environmental regulations, the effluent treatment requires attention. Anaerobic digestion is the most appropriate first-step process for spent wash treatment since biodegradable organic matter can be converted into a biofuel (biogas) by the action of microorganisms. However, the anaerobically digested effluent still contains a significant amount of sulfates, nitrates, coloring matters, heavy metals, biological oxygen demand (BOD), and biochemical oxygen demand (COD) (Mohana et al. 2007; Ravikumar et al. 2011; Mabuza et al. 2017; Kaushik et al. 2018; Malik et al. 2019a,b). In India, the existing full-scale distillery effluent treatment system includes the combination of anaerobic digestion and a two-stage extended aeration process. Thus, this huge quantity of spent wash after anaerobic digestion enters into the lagoons where it is aerated to reduce the BOD and COD and afterward the spent wash is disposed into the environment. During this treatment, the solid particles settle in the lagoons to form a sludge as a solid waste that can be used as biofertilizer because of its high nutritive value. Moreover, distilleries also produce a huge quantity of sludge per day during anaerobic digestion of spent wash, which is characterized by high organic matter, phenol, sulfate, and heavy metal along with Maillard reaction products (Singh et al. 2014; Chandra and Kumar 2017b,c; Chandra et al. 2018b). However, prior to application, distillery waste must be processed properly using the appropriate treatment method. The conventional aerobic-anaerobic secondary treatment approaches have been found ineffective to remove the color of distillery effluent due to the recalcitrant nature of melanoidins and presence of other complex co-pollutants (Jiménez et al. 2003). On disposal in the natural ecosystems, this dark brown sludge upsets natural food chains, immobilizes nutrients, affects microflora and macrofauna, and poses a risk to human health. Distillery sludge does not have farmer acceptability as its application depresses microbial activity, increases nutrient loss, and has ill-effects on crops if applied without adequate treatment or stabilization in agriculture fields. The traditional disposal methods such as open dumping and/or land filling practices by these sludge are not only increasingly expensive but impractical as open space becomes limited. Contamination of groundwater,

DOI: 10.1201/9781003029885-9

soils, as well as food resources are some of the major environmental problems that have resulted from landfilling practices of dumped waste materials. Therefore, there is an instant need for safe technology to manage such noxious distillery waste; the technologies must be ecologically viable and socially acceptable. In open environment, toxic distillery sludge is gradually converted into a soil amendment by microflora (composting) and/or macrofauna; therefore, their use singly or in combination is popular nowadays for rapid recycling of nutrients of such wastes. The product of microbial degradation, called as traditional compost, has great agronomic value. However, nitrogen immobilization and salinity effects, along with long period for conversion, are the main constraints of traditional composting. Traditional composting involves the degradation of organic waste by microorganisms under controlled conditions, in which the organic material undergoes a characteristic thermophilic stage that allows sanitization of the waste by the elimination of pathogenic microorganisms (Hemalatha 2012; Malik et al. 2019a). Thus, vermicomposting is getting enormous importance in the amelioration of severe problems associated with the disposal of large quantities of organic wastes (Mahaly et al. 2018). Several studies have revealed that vermicomposting could be an appropriate technology to transfer energy-rich organic wastes to value-added products (Singh et al. 2014; Suthar and Singh 2008). Vermicomposting is a simple biotechnological process of composting, in which certain species of earthworms is used to enhance the process of waste conversion and produce a better end product (vermicompost) (Mahaly et al. 2018). Vermicompost has no such ill-effects and is more stable and nutrient-rich. It improves the physical, chemical, and biological properties of the soil and contributes to organic enrichment (Singh et al. 2014). This chapter presents a brief overview and a bibliographic update of current knowledge and recent findings on the vermitechnology for remediation of organic and inorganic contaminants from distillery effluent and converts this hazardous waste into useful products (vermicompost). Moreover, how earthworms could be exploited to degrade this waste into vermicompost has also been discussed. Further, advantages and limitations of vermitechnology have also been illustrated.

9.2 Biology of Earthworm and Vermicompost

Earthworms, members of phylum Annelida and class Oligochaeta, are extremely important part of soil macro-fauna, especially in the upper 15–35 cm of soil where their feeding and burrowing activity affects plant growth, nutrient turnover, especially nitrogen (N) mineralization and water infiltration, soil properties, and seedling development and so have been described as "ecosystem engineers" (Figure 9.1). More than 4400 species of earthworms are distributed worldwide and occur in diverse habitats, vary greatly in size, and feed upon a variety of organic materials. Mainly they are burrowing animals and feed on dead organic matter, living bacteria, rotifers, nematodes, fungi, and other microorganisms. They also have an important impact on soil microbial communities, which in turn affects nutrient cycling and plant development through their interactions. Earthworms based on ecological and trophic niche are divided into three main categories: epigeic, anecic, and endogeic. Most suitable species for vermicomposting are epigeic because these live in organic horizons and feed primarily on decaying organic matter. Among the epigeic earthworms, *Eisenia fetida*, and *Eisenia andrei* are most commonly used in vermicomposting because both earthworms are ubiquitous with a worldwide distribution, resilient, and have wide temperature tolerance. Two tropical species, African night crawler, *Eudrilus eugeniae*, and Oriental earthworm, *Perionyx excavates*, and two temperate ones, red earthworm, *E. andrei*, and tiger earthworm, *E. fetida*, are extensively used in vermicomposting (Veeresh and Narayana, 2013). Most vermicomposting facilities and studies are using the worms *E. andrei* and *E. fetida* due to their high rate of consumption, digestion, and assimilation of organic matter, tolerance to a wide range of environmental factors, short life cycles, high reproductive rates, and endurance and resistance during handling. Earthworm hosts millions of biodegrader microbes, hydrolytic enzymes, and hormones that help in rapid decomposition of complex organic matter into vermicompost in a relatively smaller duration of 1–2 months as compared to a traditional composting method that takes nearly 5 months.

During vermicomposting, the organic wastes can be recycled into high-value products as mediated by earthworms through gut digestion, burrowing, casting, and mucus excretion. Vermicompost is a nutrient-rich, peat-like material characterized by high porosity, high water-holding capacity, and low C:N ratio (Figure 9.2). When used as an amendment for soil or other plant growth media,

FIGURE 9.1 The typical view of an earthworm.

FIGURE 9.2 A view of the developed vermicompost.

vermicompost stimulates growth, seed germination and development, flowering, and fruit production of a variety of plant species. The use of such technology will help in cost-management in agriculture, which has increased in recent years and has added to the burden of farmers in terms of chemical fertilizers and pesticides.

The mechanism of formation of vermicompost by earthworms occurs in the following steps: organic material consumed by earthworm is softened by the saliva in the mouth of the earthworms. Food in esophagus is further softened and neutralized by calcium and physical breakdown in muscular gizzard, which results in particles of size <2 μm, thereby giving an enhanced surface area for microbial processing. This finally ground material is exposed to various enzymes such as protease, amylase, lipase, cellulase, and chitinase secreted in lumen by stomach and small intestine. Moreover, microbes associated with intestine facilitate breaking down of complex biomolecules into simple compounds. Only 5–10% of the ingested material is absorbed into the tissues of worms for its growth and the rest is excreted as vermicast. The vermicast is a good organic fertilizer and soil conditioner. High-quality vermicast can be produced by worms such as the red wrigglers (*E. fetida*) as it contains humus with high levels of nutrients that has good potential for the production of organic fertilizer.

9.3 The Principles of the Composting Process

Composting is the most suitable bioconversion method for solid waste disposal and recycling of organic waste. Vermicomposting is stabilization of organic material involving the joint action of earthworms and microorganisms (whereas the other process relies solely on microbes) (Suthar and Singh, 2008; Singh et al. 2014). Earthworm accelerates the biooxidation/transformation of organic waste material into more stabilized forms by aeration and bioturbation, by their excreta, and qualitative or quantitative influence upon the telluric microflora. Soil microflora is a labile pool of nutrients, and release of the nutrients is partly regulated through grazing by the macrofauna, especially earthworms. Earthworms develop a mutualistic relationship with soil microflora, which gives them the ability to digest and convert low-quality organic matter into a nutrient-rich product. The casts remain in the soil for years together and significantly affect immobilization or release of nutrients in the ecosystem. Soil microflora is a labile pool of nutrients, and release of the nutrients is partly regulated through grazing by the macrofauna, especially earthworms. Compared to thermal composting, vermicomposting with earthworms often produces a product with a lower mass, lower processing time, and humus content; phytotoxicity is less likely; more N is released; fertilizer value is usually greater; and an additional product (earthworms), which can have other uses, is produced (Figure 9.3).

9.4 Case Studies of Vermicomposting and Vermifiltration of Distillery Waste

Bioremediation of distillery sludge into soil-enriching material through vermicomposting with the help of *E. fetida* has been reported by Singh et al. (2014). In this study, distillery sludge was mixed with a complementary waste, cattle dung, and subjected to vermicomposting with and without *E. fetida* in a different ratio. It was observed that nitrogen, phosphorus, sodium, and pH increased during vermicomposting but decreased in the products without earthworm, and there was increase in the contents of transition metals in the products of both the techniques. This study indicated that the compost prepared after mixing the distillery sludge with cattle dung can be used in the fields without any ill-effects on soil or plants. Vermicomposting of distillery sludge waste with tea leaf residues (TLR) is a most effective method for bioconversion of these organic wastes into manure and probably minimizing the environmental pollution caused by their disposal (Mahaly et al. 2018). In this approach, distillery spent wash and TLR in different combinations involve the earthworm, *E. fetida*. Further, periodical physicochemical analysis of different combinations of vermicompost

FIGURE 9.3 A view of the vermicomposting processes.

was analyzed in vermicompost samples taken on day 0, 15, 30, and 45. It was observed that the values of pH, total nitrogen, total phosphorus, total potassium, and available nitrogen, phosphorus, potassium, total calcium, and total magnesium were increased with a declining trend in electrical conductivity, total organic carbon (TOC), and carbon to nitrogen (C/N) ratio from its initial value. Finally, the solid wastes were successfully transformed into nutrient-rich vermicompost. The feasibility of vermicomposting technology to stabilize the distillery industry sludge mixed with cow dung in different proportions was demonstrated using composting earthworm *Perionyx excavatus* in 90 days (Suthar and Singh 2008). Further, the vermitreated sludge was evaluated for determination of different physicochemical parameters and all vermibeds expressed a significant decrease in pH, organic carbon contents, and an increase in total N, available P as well as exchangeable potassium (K: 95.4–182.5%), calcium (Ca: 45.9–115.6%), and magnesium contents (Mg:13.2–58.6%). Data suggested that inoculated earthworms maximize the decomposition and mineralization rate if sludge was used with appropriate bulking material for earthworm feed. Vermicomposting also caused a significant reduction in a total concentration of zinc (Zn), iron (Fe), manganese (Mn), and copper (Cu) in sludge. The greater values of BCFs indicated the capability of earthworms to accumulate a considerable amount of metals in their tissues from the substrate. Besides the use of vermicomposting, vermifiltration technology has also been applied for biological treatment of distillery wastewater (Manyuchi et al. 2018). During vermifiltration, earthworms are used as the biological filtration media for removal of biocontaminants present in wastewater. Authors used 10 kg of *Eisenia fetida* earthworms as the vermifiltration media in vermifiltration bed over a 40-h period cycle. The COD, BOD, total Kjeldahl nitrogen (TKN), total soluble solids (TSS), and total dissolved solids (TDS) were significantly reduced by more than 90% during the 40-h vermifiltration process. The treated distillery effluent can be used for irrigation purposes. In addition, the vermicompost, a by-product of the vermifiltration process, had a nitrogen, phosphorous, and potassium composition of 1.87%, 0.87%, and 0.66%, respectively. The major physicochemical characteristics of compost produced by vermicomposting of distillery effluent are described in **Table 9.1**.

TABLE 9.1

A Typical Physicochemical Characteristics of Raw Distillery Effluent and Vermicompost Prepared from Distillery Effluent (Manyuchi et al. 2018)

S. No.	Parameters	Raw Distillery Effluent	Vermicompost Prepared from Distillery Effluent
1.	pH	3.7 ± 0.2	7.3 ± 0.1
2.	TKN	$1,173.3 \pm 25.2$	60.0 ± 5.0
3.	BOD	$25,466.7 \pm 814.5$	2266.7 ± 208.2
4.	TDS	$65,116.7 \pm 600.7$	52766.7 ± 971.3
5.	TSS	3466.7 ± 450.9	263.3 ± 35.1
6.	COD	$92,500.0 \pm 500.0$	9783.3 ± 464.6

All parameters are in milligrams per liter except for pH. TKN: total Kjeldahl nitrogen; BOD: biochemical oxygen demand; COD: chemical oxygen demand; TDS: total dissolved solids; TSS: total soluble solids

9.5 Advantages and Limitations of Vermitechnology

Vermicomposting, a more appropriate and an efficient technology to convert industrial waste to valuable community resources at low-input basis, has made tremendous gains in market acceptance in recent years. Its primary advantage is that it is eco-friendly than traditional composting practices. In addition, the other advantages of vermicomposting, in comparison with classical composting methods, can be summarized as follows:

i. Vermicomposting, or worm farming, has been a part of organic and natural gardening methods for years.
ii. Vermicomposting requires a low temperature to be maintained in the compost operation, so the worms do not lose moisture and dry out. The normal temperature for vermicomposting is between 55° and 80°F.
iii. Less production of toxic gases.

Although vermicomposting is environment-friendly technology for ecological restoration of industrial-waste-contaminated sites, this technology has several disadvantages and limitations that could create a hinder implementation of the strategy:

i. Vermicomposting requires more space than regular composting.
ii. Vermicomposting requires greater care than regular composting methods. The worms used in vermicomposting require care, food, heat, and moisture at proper levels to remain healthy and create the compost material.
iii. It is more expensive to set up than regular compost piles or batch composters.
iv. Vermicomposting needs special materials to start, such as plastic or metal containers and red worms.
v. Vermicomposting also requires lime to stabilize the acid levels created by the waste in the new soil.
vi. Vermicomposting requires waste to be applied in thin layers because of temperature concerns. If you apply too much waste to the bin in a short amount of time, the temperature will rise too fast and kill off the worms.

10

Sequential Treatment: A Novel Approach for Biodegradation and Detoxification of Distillery Effluent for Environmental Safety

10.1 Introduction

The effluent generated from distillery industries during ethanol production contains toxic refractory chemical compounds, including nitrogenous compounds (melanoidins), phenolics, endocrine-disrupting chemicals (EDCs), and heavy metals, that are significantly deteriorating the quality of the environment as well as pose a great threat to human health. Effluent from the distillery industry has high biological demand (BOD) and chemical oxygen demand (COD), which would cause the destruction of microorganisms that are useful in biodegradation (Arimi 2017; Ahmed et al. 2020). Distilleries are conducted toward the treatment of their toxic effluents using various conventional as well as advanced physicochemical technologies in an attempt to meet the requirements of the increasingly stringent regulations despite their failure to produce reusable water. These technologies, however, have some drawbacks such as high operating and equipment cost, operational difficulty, and excess use of chemicals, they are inefficient for color removal and pollutants detoxification, and sometimes they generate a huge amount of sludge with subsequent disposal problems and occasionally form hazardous by-products/secondary pollutants (Rani et al. 2013; Manyuchi et al. 2018; Reis et al. 2019). These treatment processes are most commonly used as a pre- or posttreatment step in addition to the main biological treatment or are directly used for spent wash. Considering the limitation of different effluent treatment technologies, there is no individual technique that can be used to treat/mineralize and/or detoxify distillery effluent (Apollo and Aoyi 2016; Miyata et al. 2000). Thus, inadequate treatment of toxic chemicals makes the treatment of distillery effluent a major challenge with several negative impacts such as inhibition of aquatic photosynthesis, depletion of dissolved oxygen, chronic toxicity, and so on. Hence, there is a need to establish a comprehensive effluent treatment approach involving sequential/phase separation or combined (a combination of the various techniques in a hybrid method) treatment technologies. Numerous scientific reports indicated that treatment of distillery effluent by sequential/phase separation method or combined/hybrid methods has been a promising approach for degradation and detoxification of distillery effluent due to the less use of energy and high removal of nutrients and other chemicals like heavy metals from the effluent. This chapter deals with the existing conventional, sequential, and hybrid treatment approaches for safe disposal of distillery wastewater into the environment.

10.2 Conventional Sequential Treatment

The typical treatment sequence is screening or equalization, followed by biomethanation. The highly concentrated effluents are made profitable on-site by anaerobic digestion, which produces methane while removing more than 80% of the organic load (Lettinga 1995; Kaushik et al. 2000; Mabuza et al. 2017). The biomethanated effluent is occasionally subjected to a single- or two-stage aerobic treatment through the activated sludge, trickling filters, or even a second stage of anaerobic treatment in anaerobic lagoons. After the anaerobic-aerobic treatment, final problematic wastewater is generated as it still contains an organic load that is too high for release into the environment (Malik et al. 2019a,b). Furthermore, it is not biodegradable and so it is not suitable for treatment in municipal sewage plants. Moreover, the dark brown color of such wastewater remains an issue for its disposal (Pant and Adholeya 2009b). The colored content is generally assumed to be melanoidin with a high molecular weight produced during Maillard reactions in sugar processing. In order to increase the biodegradation ability of the process, a two-stage sequential/phase separation/sequential method has demonstrated to be an efficient approach for bioremediation of distillery effluent.

10.3 Sequential Treatment Approaches

Investigation in implementing a hybrid method of treating the distillery effluent has gained its soundness rather than individual treatment. For instance, Ghosh et al. (2002) treated distillery effluent in a two-stage bioreactor by using *Pseudomonas putida* U followed by *Aeromonas* sp. strain EMa. At the first stage, *P. putida* decreased the color and COD of distillery effluent up to 60% and 44.4%, respectively, whereas in the second stage, *Aeromonas* sp. strain EMa reduced the effluent COD up to about 44.4%. Kumar and Chandra (2004) successfully treated post-methanated distillery effluent (PMDE) in two-stage treatment process involving the transformation of recalcitrant coloring components of the PMDE by aerobic bacterium *Bacillus thuringiensis* followed by a subsequent decline of a remaining load of pollutants by a macrophyte *Spirodela polyrrhiza* Schleiden. A similar biphasic (two-step) treatment of the PMDE was carried out in a constructed wetland (CW) with *B. thuringiensis* followed by *Typha angustata* L. by Chandra et al. (2008), which resulted in COD, BOD, and color reduction after 7 days of treatment. The results suggested that the bacterial pretreatment of PMDE integrated with CW improved the treatment process of PMDE and promote safe disposal of hazardous distillery waste. Pant and Adholeya (2009b) examined the removal of nitrogen from anaerobically digested wastewater in a two-stage sequential treatment processes. In the first step, the wastewater was treated in a hydroponic-based system using two plant species, namely, *Vetiveria zizanioides* and *Phragmites kharka*, to reduce the total Kjeldahl nitrogen content of the effluent. Roots of these plants showed profuse growth on effluent. Nitrogen removal to the tune of 84% was achieved. When this hydroponically treated effluent was subjected for treatment by fungal isolates, 86.33% decolorization was obtained with *Pleurotus florida* Eger EM1303 followed by *Aspergillus flavus* TERIDB9 (74.67%), with a significant reduction in COD as well. The advantage of using two-step sequential treatments as suggested is using plant-based system for removal of nitrogen from cane molasses-based distillery effluent in the first step in the reactor and subsequent treatment of effluent by fungus for effective removal of the recalcitrant color causing organic compounds. Authors recommended distillery effluent treatment without the need of high dilutions and addition of supplementary carbon sources. Distillery spent wash treated by coagulation with polyaluminium chloride (PAC) followed by biological treatment with fungal sequencing batch aerobic reactor (FSBAR) resulted in 87% decolorization as has been reported by Singh and Dikshit (2012). To achieve further decolorization, ozonation of treated distillery spent wash was carried out at different ozone doses and contact times. The optimum ozone dose was found to be 4.75 g/L at an application rate of 3.8 g/h for 30 min with corresponding decolorization being 66%. The overall decolorization obtained by the combined treatment (PAC treatment, FSBAR, and ozonation) was 96% and the total COD removal was 81%. The potential use of fungi (*Cladosporium cladosporioides*) and cyanobacteria (*Phormidium valdernium*) for treatment of DW in a two-stage sequential step was also reported by Ravikumar and Kartik (2015). A maximum 68.5% decolorization and 81.37% COD reduction have been achieved in the first-stage bioreactor during the batch experiment. Further, the spent wash from bioreactor was treated with cyanobacteria in the second stage and resulted in COD reduction (3652 mg/L) of 89.5% and 92.7% decolorization, respectively. Authors recommended that sequential treatment using the combination of fungi and cyanobacteria resulted in better decolorization and degradation of spent wash. Combination of wetland treatment technology after bacterial degradation offers an excellent system for the elimination of color from PMDE and reduction of BOD, COD, TDS, and heavy metals for safe disposal. Chandra et al. (2012) reported 96% and 94.5% reduction in COD and BOD values in a two-step sequential treatment of PMDE by bacteria and wetland plant *Phragmites communis*. Kumar (2018 b) developed a two-step

FIGURE 10.1 A developed zigzag-shaped pilot-scale horizontal subsurface flow-constructed wetland plant treatment system vegetated with *Phragmites communis* (Kumar 2018b).

sequential treatment system for degradation and detoxification of post-methanated distillery effluent as shown in Figure 10.1.

At the first stage, PMDE was treated by a developed bacterial consortium comprising *Klebsiella pneumoniae*, *Salmonella enterica*, *Enterobacter aerogenes*, and *Enterobacter cloaceae*, while at the second step bacteria-treated PMDE was circuited into a horizontal subsurface flow-constructed wetland (HSSF-CW) plant treatment system vegetated with *P. communis* plants (Figure 10.2). Author found that bacterial consortium-treated sample of PMDE obtained after 168 h incubation showed a reduction

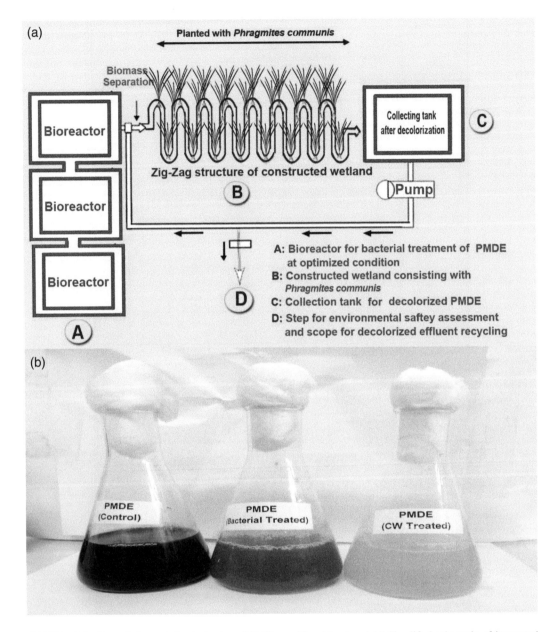

FIGURE 10.2 Degradation and decolorization of distillery effluent treatment (a) Simplified schematic of integrated bacterial and subsurface horizontal flow constructed wetland treatment system for post-methanated distillery effluent treatment (PMDE) (b) Erlenmeyer flask shows decolorization of post-methanated distillery effluent after bacterial and constructed wetland treatment (Kumar 2018b).

in all physicochemical parameters compared to control. Afterward, the supernatant of bacteria-treated PMDE was integrated into HSSF-CW plant treatment system with different flow rate and hydraulic retention time to obtain the more decolorization and reduction of physicochemical parameters. The constructed wetland plant rhizosphere treated PMDE sample has also shown a reduction of BOD COD, TDS, SS, VS, TN, chloride, sulfate, phenol, and various heavy metals such as Fe, Zn, Ni, MN, Pb, Cu, and Cd. This study indicated that during the bacterial treatment of PMDE, organic compounds and metal complexes get biotransformed into such forms that become easily bioavailable to wetland plants rhizospheric microflora and plant root. The decolorization of PMDE is a combined phenomenon of bacterial transformations followed by the bacteria-assisted wetland plants rhizospheric treatment, which grows in the vicinity of plant roots and supported the bioaccumulation of toxic metals by wetland plants. The growing bacterial communities helped in degradation of toxic wastes and the plant accumulates the heavy metals in its tissues. Both of them played a major role in the bioremediation of PMDE. Therefore, all the values of BOD, COD, phenol, sulfate, and heavy metals were reduced significantly.

Kaushik et al. (2010) treated distillery spent wash in a three-stage bioreactor using *Emericella nidulans* var. lata, *Neurospora intermedia*, and *Bacillus* sp. followed by bacteria. The treated effluent showed a significant reduction in color (82%) and COD (93%) after 30 h. Denaturing gradient gel electrophoresis of 16S rDNA and 18S rDNA gene sequences amplified from DNA and isolated from the reactor communities indicated the presence of other organisms besides those introduced initially. The microbial communities were able to carry out bioremediation of distillery effluent and produce a discharge that conforms to safety standards. Significant reduction in color (82%) and COD (93%) was achieved when microbes were applied in a stepwise manner. A novel three-step sequential/combinatorial/hybrid technology for the treatment of four molasses-based raw industrial effluents, varying in their COD, color, and turbidity, is reported by Verma and his colleagues in 2011. Sequential steps involved in this treatment are (1) sonication of the effluents, (2) whole-fungal treatment of these by a ligninolytic marine fungus, and (3) biosorption of the residual color with heat-inactivated biomass of the same fungus. Sonication reduced the foul odor and turbidity of the effluents. It increased their biodegradability of the effluents in the second stage of treatment. Laccase production in the presence of all the four effluents was directly correlated with their decolorization. After the third step, a reduction of 60–80% in color, 50–70% in COD, and 60–70% in total phenolics were achieved. Comparative mass and nuclear magnetic resonance spectra indicated increasing degradation of the effluent components after each stage. Toxicity (LC_{50} values) against *Artemia* larvae was reduced by two- to fivefolds (Verma et al. 2011).

10.4 Hybrid Treatment Approaches

The combination of two or more treatment methods in single stage results in the creation of more economical and effective hybrid method. Numerous scientific reports indicated that the use of a hybrid technique by using bacteria, fungi, yeast, and plant or their combinatorial systems is more successful than the individual one. Distillery wastewater was treated using ozone as a chemical oxidant in combination with conventional aerobic oxidation. The combination of chemical and biological processes led to greater destruction of organic contaminants present in the effluent (Sangave et al. 2007). Accordingly, in the first phase of work, the biological degradation or aerobic oxidation by microorganisms and the chemical oxidation by ozone have been studied separately, with an aim of quantifying the color and COD removal efficiencies. In the second phase, the combined processes consisting of ozonation followed by aerobic oxidation were carried out to establish the COD removal efficiency achieved by these processes in series. In the third phase, ozonation was used as the method of choice to remove the colored products from the biotreated effluent sample (thus as a post-aerobic treatment step), which would, in turn, enhance the effectiveness of the overall treatment scheme for distillery spent wash. Mabuza et al. (2017) investigated the synergy of integrated anaerobic digestion and photodegradation using hybrid photocatalyst for molasses wastewater treatment. A hybrid photocatalyst consisting of titanium dioxide (TiO_2) and zinc oxide (ZnO) was used for photocatalytic degradation. Biodegradation at thermophilic conditions in the bioreactor achieved high total organic carbon (TOC) and COD reductions of 80% and 90%, respectively, but with an increased color intensity. Contrastingly, UV photodegradation achieved a high color reduction of 92% with an insignificant 6% TOC reduction after 30 min of irradiation. During photodegradation, the mineralization of the recalcitrant organic compounds led to the color disappearance. In this system, the anaerobic digestion will first remove the high organic content followed by color removal by the UV process. This integrated process, however, needs to be optimized to determine the best operational and combinational factors. Evaluating the overall energy demand of such a system is also important in determining whether the bioenergy produced from the anaerobic digestion process can be used to supplement the energy requirement of the integrated system. Reis et al. (2019) introduced a novel alternative to the treatment of vinasse promoting the reduction in COD levels, phenolic compounds, and its mineral content through the coupling of ozone treatment, anaerobic digestion, and the aerobic growth of fungi *Mucor circinelloides*. The ozone treatment can remove about 30% of the total COD and deplete the concentration of phenolic compounds, while (posttreatment) anaerobic digestion produces biogas and generates vinasse digestate, which is less biorecalcitrant than the raw vinasse. The aerobic fungal growth generates oleaginous fungal biomass and promotes over 80% of Kjeldahl nitrogen in the vinasse. If vinasse were treated following the sequence of anaerobic digestion, aerobic fungal growth, and ozone treatment, the effluent would have about 95% of the COD decreased, complete removal of phenolic compounds, and over 80% of Kjeldahl nitrogen.

11

Microbial Fuel Cell Technology in Distillery Wastewater Treatment and Bioelectricity Generation

11.1 Introduction

The alcohol-producing molasses-based distillery industry is responsible for an extensive list of environmental impacts. The main damages caused by the distillery industry to the environment, however, are those resulting from the discharge of large volume of very complex high-strength color effluents as a by-product into the aquatic bodies (Wilk and Krzywonos 2020). Due to the intense color of the effluent, even after the secondary treatment, the safe disposal of effluent is always a major concern for environmentalists (Hati et al. 2004; Kumar and Chopra 2012; Reis and Hu 2017). If the wastewater from effluent from distilleries is discharged accidentally into nearby irrigation canals, it causes pollution in seas, lakes, and rivers by lowering pH, increasing organic and inorganic burden, depleting oxygen, discoloring water, and destroying aquatic life (Ramakritinan et al., 2005). To relate the severity, it is stated that the color effluent requires 1000 times dilution to suppress the effect on the receiving water bodies. Therefore, the Central Pollution Control Board (CPCB), Government of India, insisted that all distilleries achieve zero liquid discharge (ZLD). To achieve ZLD, distilleries follow numerous techniques like evaporation followed by incineration or anaerobic digestion followed by reverse osmosis (RO) process as an effluent treatment processes (ETP) (Sankaran et al. 2015). Evaporation followed by incineration is costly and energy-intensive since the wastewater contains only 9–11% of solids. Anaerobic digestion is a widely practiced process that effectively removes 65–70% of COD, 80–90% of the BOD, while generating biogas simultaneously (Sankaran et al., 2014; Bhoite and Vaidya, 2018a). The biogas is further used to generate electricity, thereby meeting the power requirement of the distilleries. Anaerobic digestion is followed by RO process to separate the clean water from the effluent and recycle it to the production unit. However, the RO process has the following disadvantages: (i) high pollution load of influent (ii) need of high pressure for the operation (iii) consumption of power (iv) high maintenance cost and (v) frequent replacement of membranes of system. Thus, researchers are looking for suitable and cost-effective management options for distillery effluent, such as reuse of effluent for the electricity generation (Ha et al. 2012; Naina Mohamed et al. 2018). Distillery effluent is an important potential sources for electricity generation using microbial fuel cells (MFCs) because it has a high content of the organic waste and can be easily degraded (Sonawane et al., 2014; Ghosh Ray and Ghangrekar 2015). MFCs are bioelectrochemical devices that couple organic carbon removal from wastewater/effluent and simultaneous production of electricity. The use of MFCs as an alternative source for power generation is considered as a reliable, clean, and efficient process, which utilizes renewable methods and does not generate any toxic by-product (Ha et al. 2012; Quan et al. 2014). Therefore, in recent years, MFCs have shown to be a potent technology for recovery and in situ conversion of chemical energy into electricity. This chapter aims to provide a brief introduction to MFCs technology and its principle and mechanism in wastewater treatment and bioelectricity generation. Moreover, this chapter gives an insight into the major challenges holding back the application and the scaling-up of the MFCs technology, including the thermodynamic limits, heavy metals biotoxicity, pH imbalance, and membrane biofouling.

DOI: 10.1201/9781003029885-11

11.2 Principles and Mechanism of Microbial Fuel Cells Technology

MFC, similar to any other battery or fuel cells, is attracting wide attention due to its intended use to recover chemical energy in the form of electricity. An MFC is a system where microbes convert chemical energy produced by the oxidation of organic compounds into ATP by sequential reactions in which electrons are transferred to a terminal electron acceptor to generate electrical energy. In their simplest form, MFCs are consisting of two electrodes, an anode and a cathode, which are generally separated by a proton exchange membrane (PEM) to avoid the migration of electrolytes from one chamber to the other. Microbes reside in the anode compartment where they metabolize carbon sources or substrates to release electrons and protons. In principle, anaerobic microbes forming an electroactive biofilm are used as catalysts in MFCs to harvest energy from the organic matter existing in the effluent. Microorganisms that oxidize organic compounds and transfer electrons to the anodes surface of MFCs are called electricians, exoelectrogens, electroactive microorganisms, or active biocatalyst (Chaturvedi and Verma 2016). Exoelectrogens can be defined as prokaryotes that have the capability of interacting with charged conductive electrode surfaces, using them as either donors or acceptors of electrons, with both Gram-positive and Gram-negative electricigens. This electron transport is carried out via extracellular electron transport (EET), in which several mechanisms are employed. The electrons produced during oxidation of organic compounds flow from the anode to the cathode via external electric circuit to produce current, thus electricity is generated. while the protons released from the anaerobic respiration diffuse through the PEM. In the cathode, electron acceptors such as oxygen or hexacynoferrate and acidic permanganate react with electrons and protons to produce reduced compounds such as water. The reduction of oxygen to water is the most common cathodic reaction. To accelerate the oxygen reduction on the surface of the cathode, platinum is commonly used because of its excellent catalytic ability. Many compounds other than oxygen could also be used as terminal electron acceptors in the cathode, such as nitrate, sulfate, iron, manganese, selenate, arsenate, urinate, fumarate, carbon dioxide, and hexavalent chromium (Cao et al. 2019). Any compound with a comparable or higher redox potential than oxygen can be reduced at the cathode; thus, elements like permanganate, ferricyanide, nitrate, persulfate, dye molecules, and most importantly heavy metals were used and proven to be efficient electron acceptors in the cathode of MFCs. This is due to the fact that the reduction of these compounds is thermodynamically favorable that makes the flow of electrons from the anode to the cathode occur spontaneously and without any external power consumption. Oxygen in the anode chamber will inhibit the production of electricity; thus, a pragmatic system must be designed to keep the bacteria separated from oxygen (anaerobic chamber for anodic reaction). This provides a potential approach for wastewater treatment using biocathode due to its variety of terminal electron acceptors such as recalcitrant distillery wastes. There are three possible ways used by microorganisms to transfer electrons to the anode electrode, including (a) direct contact; (b) pili/conductive wires; and (c) redox mediators or electron shuttle. It should be noted that microorganisms can use one or the combination of multiple mechanisms at a time for electron transport. In general, the process of organic matter oxidation at the anode is coupled to the reduction of oxygen at the cathode. A typical design of MFCs is illustrated in Figure 11.1. The output of MFC depends upon a number of parameters such as its configuration, type of substrate, its concentration, microorganism used, catalyst, materials used in cathode and anode, and suitable membrane, which play an important role in its performance.

11.3 Case Studies of Distillery Wastewater Treatment Using Microbial Fuel Cells Technology

MFCs have been nominated as new alternatives and novel opportunities that are able to convert biodegradable organic matters as substrates into green electricity with the aim of different types of active microorganisms as active biocatalysts. MFCs is an efficient technology used for electricity recovery and pollutants removal from wastewater. Recently, molasses distillery wastewater has been examined as an organic fuel for electricity production in mesophilic MFCs (Zhang et al. 2009; Naina Mohamed et al. 2018). Zhang et al. (2009) studied the treatment of high-strength molasses distillery wastewater in an

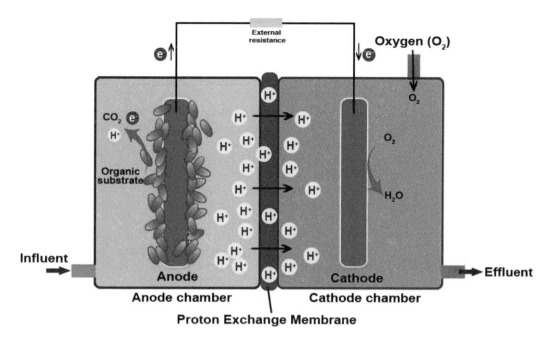

FIGURE 11.1 The typical structure of a developed microbial fuel cell (MFC).

integrated UASB-MFC system of 2.2 L and found 53% removal of COD and much higher power generation of 1.4 W/m^2 in MFC receiving effluent from anaerobic process. Simultaneous electricity generation and distillery wastewater treatment using a bacteroidetes-dominant thermophilic MFC has been investigated by Ha et al. (2012). In this study, thermophilic MFC, which require less energy for cooling the distillery wastewater, can achieve high efficiency for electricity generation and also reduce sulfate along with oxidizing complex organic substrates. The generated current density (2.3 A/m^2) and power density (up to 1.0 W/m^2) were higher than previous wastewater-treating MFC. Further, bacterial diversity analyzed by pyrosequencing of the 16S ribosomal RNA gene revealed that uncultured bacteroidetes thermophiles found dominant in the thermophilic MFC fed with distillery wastewater. Sonawane et al. (2014) treated domestic and distillery wastewater in high-surface MFCs. In this study, a high-surface MFCs anode, arranged in a double-air cathode MFC configuration with 6 U internal resistance, is constructed by interlacing carbon yarn with stainless steel. The MFCs produce maximum power densities of 621 and 364 mW/m^2 for domestic and distillery wastewater, respectively. The COD removal was achieved up to 68% and 58% with a columbic efficiency of 47% and 27% for domestic and distillery wastewater, respectively. The rapid decrease of current with time in the distillery wastewater MFC suggests that distillery wastewater alone does not support efficient electrochemically active microbial communities. Preformentation coupled with halfway anode aeration is a feasible strategy to enhance power generation and pollutants removal from the cassava wastewater in MFC. Quan et al. (2014) investigated the effects of pretreatment and anode aeration on electricity recovery and pollutants removal from the cassava alcohol wastewater in MFCs. Different pretreatment methods such as solid-liquid separation, ultrasonication, preformentation, and anode-aeration modes were explored in MFCs aimed to enhance the efficiency of power generation and pollutants removal. Authors achieved a maximum power density of 437.13 mW/m^2 and COD removal efficiency of 62.5% using the preformented wastewater, 150% and 20% higher than the un-pretreated control. In this system, in situ anode aeration promoted the hydrolysis of organic matter and production of volatile fatty acids in the raw wastewater, which led to a reduced operation time and an obvious increase in COD removal and power density. Similarly, for enhancing organic matter removal from cereal-based distillery wastewater, a two-stage treatment consisting of fermentation by *Aspergillus awamori* followed by MFC has been examined by Ghosh Ray and Ghangrekar (2015). Overall, 99% reduction in COD and suspended solids and threefold improvements in power generation in MFC were accomplished in this system. Authors recommended that fungal pretreatment and further treatment in MFC offered a technological solution for the treatment of the distillery wastewater to solve the critical problem of water pollution. Naina Mohamed et al. (2018) studied the effect of feed pH and buffering conditions on electricity production and treatment efficiency using distillery wastewater as a substrate in MFC. In the MFC system, the anodic chamber was operated with diluted distillery wastewater at various pH between 5.4 and 10, while the cathode chamber was maintained at pH 7.5. The MFC peak power density of 168 mW/m^2 (580 mA/m^2) with COD, color and total dissolved solids (TDS) removal efficiency of 68.4%, 26.4%, and 15.4%, respectively, was achieved at pH 8. This study concluded that the optimum pH (8) and borate buffer enhanced the power generation and treatment efficiency from distillery wastewater in the MFC. Nayak et al. (2018) developed an innovative mixotrophic approach of distillery spent wash with sewage wastewater for biodegradation and bioelectricity generation using MFC. This study demonstrated that proper dilution of distillery spent wash with sewage wastewater may lead to an efficient wastewater remediation and energy production. H-type MFC has been used to treat distillery spent wash diluted with sewage wastewater at different mixing ratio. Mixture of distillery spent wash and sewage wastewater was used as substrate in anaerobic environment. The cathodic compartment was filled with BBM media and microalgae (*Scenedesmus abundans*). CO$_2$ generated during breakdown of organic substrates by anaerobes present in distillery spent wash was utilized by microalga *S. abundans* for photosynthesis. A consortium of two bacteria (*Pseudomonas aeruginosa* and *Bacillus cereus*) was used for the 21 days period of spent wash treatment. Significant reduction in COD on the order of 66–78.66% was observed. At 50:50 ratio of spent wash and sewage wastewater, TDS of 39.66% and total suspended solid (TSS) of 97% were removed. A maximum power density of 836.81 mW/m^2 and open-circuit voltage of 745.13 mV were obtained along with biomass yield of 0.74 g/L per day with above ratio of distillery spent wash and sewage wastewater.

11.4 Advantages and Limitations of MFCs Technology in Distillery Effluent Treatment

The success of any technology depends upon its commercialization when it is marketed in huge amounts and used by a large number of people. Since MFC deals with the production of electricity by employing waste materials, its commercialization will offer several advantages:

1. Among the next-generation energy sources, MFC is attracting wide attention due to its intended use to recover energy in the form of electricity. Production of low-cost electricity from waste materials.
2. The electricity will be produced round the year since waste and xenobiotics are readily available.
3. People would be able to produce electricity in their homes.
4. This technology will be helpful for the people living in poor countries such as Africa where huge infrastructure required for a set of energy production plants is not available.
5. MFC will lead to clean up of wastes and xenobiotics. So, it can be used as an alternative method for bioremediation.
6. Direct production of electricity out of substrates, enabling to be operated efficiently at an ambient temperature, and expanding the diversity of fuels used as energy requirements are some of the most praiseworthy advantages of MFC.

MFC technology has been improved significantly in the recent decades. However, it has encountered several challenges in scale-up and practical application, such as turbulence in each compartment and membrane resistance in the proton transportation process:

1. The high cost of platinum is a major limitation to MFC application and economic viability.
2. Practical application of MFC technology has not been realized, because of great challenges in cost, system development, and energy recovery.

3. The construction and analysis of MFCs requires knowledge of different scientific and engineering fields, ranging from microbiology and electrochemistry to materials and environmental engineering.

4. Coupled with those, MFCs have confronted two bottlenecks in power generation. More than a specific value of substrate concentration, the power generation, will be prevented.

5. MFC output is restricted while high internal resistance utilizes considerable amount of power production in MFC.

The MFC technology, although still at its infancy, might bring in new opportunities because of its many unique features. To find out whether the envisioned advantages of this technology can be ultimately achieved, we need to reexamine the challenges and feasibility of this technology, and to think about how to recalibrate the role of this technology in a future paradigm of sustainable wastewater treatment. Describing MFC systems therefore involves an understanding of these different scientific and engineering principles.

12

Biodiesel Production from Distillery Waste: An Efficient Technique to Convert Waste to Biodiesel

12.1 Introduction

Biodiesel has received a lot of interest in recent years due to declining world oil reserves and rapidly increasing world fuel consumption (85 million barrels of liquid fuel per day in 2006 and projected to increase to 107 million barrels of liquid fuel per day in 2030), which have resulted in increasing price of petroleum-based fuels. In addition to being renewable and biodegradable, biodiesel could provide displacement of imported petroleum-based fuels, similar energy density to petro-diesel and higher flash point, inherent lubricity, and reduction of most exhaust emissions (Kumar and Thakur 2020a,b). These advantages make biodiesel a promising alternative energy carrier. Ethanol production from molasses-based fermentation and distillery industries has been attracting worldwide because of the high demand due to the availability of limited non-renewable energy resources. The ethanol primarily serves as a renewable fuel blending with gasoline and diesel to increase the energy security in developed as well as developing countries, including India (Bhoite and Vaidya 2018a; Ziaei-Rad et al. 2020). It also finds its usage in the manufacturing of beverages. The production of ethanol is expected to continue to grow over the years because of Indian Government regulation to blend 5–10% of ethanol with petrol. More than 397 ethanol distilleries are functioning in India producing approximately 3.25×10^{10} L of ethanol using sugarcane molasses as raw materials, which are by-products of the sugar industry. Meanwhile, the rapid boom of the distillery industry over the past few decades also disappointingly brought significant environmental burdens due to the discharge of heavily polluted effluent (Zhang et al., 2017; Takle et al., 2018; Ahmed et al., 2020). Spent wash, also known as raw effluent, the liquid fraction generated from the rectification and distillation operations of ethanol, is a dark-colored waste, characterized by high organic matter content with a significant amount of nutrients, BOD (biological oxygen demand), COD (chemical oxygen demand), TSS (total suspended solids), and TDS (total dissolved solids), which has high pollution potential (Prajapati and Chaudhari 2015; Chuppa-Tostain et al. 2018). The disposal of untreated or partly treated effluent shows adverse effects on both flora and fauna; its discharge to the land alters physical and chemical properties of the soil, thus reducing the fertility of agricultural land for crop production and its discharge to the water bodies may result in eutrophication, affecting the aquatic life and making the water unfit for drinking (Ramakritinan et al. 2005; Alves et al. 2015; Chowdhary et al. 2018a). Thus, the challenges for safe disposal of distillery effluent cannot be ignored (Chandra et al. 2009). Environmentalists, industrialists, and government are looking for efficient, effective, and long-lasting solutions for effluent treatment, recycling, and management. The most frequently biological-based treatment processes adopted for distillery effluent treatment at common effluent treatment plants (CETPs) are the upflow anaerobic sludge blanket (UASB) and activated sludge process (ASP) (Sales et al. 1987; Ahansazan et al. 2014). Distillery sludge is an abundant organic solid waste or by-product generated in CETP facilities after primary and/or secondary treatment processes. A typical CETP having an ASP produces two main types of sludge: primary sludge, collected at the bottom of the primary clarifier, consists of a high portion of organic matters, and the secondary sludge, known also as activated sludge, composed mainly of the mixed microbial community most of which are heterotrophic bacteria, and suspended solids produced during the aerobic biological treatment, is collected in the secondary clarifier (Ahansazan et al. 2014; Kumar and Thakur 2020b). The yield of fatty acid methyl esters (FAME) from

DOI: 10.1201/9781003029885-12

primary sludge is greater than that from activated sludge. In the wastewater treatment plant process, primary sludge is a result of the capture of suspended solids and organics in the primary treatment process through gravitational sedimentation, typically by a primary clarifier. The secondary treatment process uses microorganisms to consume the organic matter in the wastewater. ASP is a potential source of lipids containing sludge solid. Lipids could be an unexploited source of cheap and readily available feedstock for biodiesel production. The lipid extraction from the raw sludge requires a huge amount of organic solvent and large vessels with stirring and heating systems. Solvent extraction especially using a mixture of chloroform and methanol is still the main extraction method used by researchers due to its simplicity and is relatively inexpensive requiring almost no investment for equipment and gives higher yields for the extraction of lipid. There are also several modifications for this lipid extraction process, including the addition of water either to the solvent mixture or to dry samples when extracting lipid, sample sonication, increasing the proportion of methanol in the extraction solvent, and extraction from wet samples. The volumes of distillery sludge are increasing all over the world and the amount of sludge is expected to increase in the future due to increasing industrialization. The sludge from distillery industries consists of organic and inorganic substances (Chandra and Kumar 2017b; Kumar and Chandra 2020b). Several studies have investigated biodiesel production using a two-step process from primary and secondary sludge. The chemical composition of inorganic substances commonly presented in distillery sludge includes the following elements and their chemical compounds: nitrogen, ammonia, sulfides, sodium, sulfates, phosphate, calcium compounds, as well as a high content of various heavy metals (Suthar and Singh 2008; Mahaly et al. 2018). Distillery sludge contains a significant amount of organic matter that is predominantly lipids, proteins, and carbohydrates. The use of sludge as a fertilizer is restricted in many countries of the world due to bad odor, the presence of heavy metals, and recalcitrant toxic substances, while the sludge incineration results in emissions of toxic heavy metals and dioxins into the environment. Moreover, distillery sludge in India cannot be disposed to landfills due to the high chemical and water content that drains away from the materials and enters the soil and groundwater. Besides, traditional treatment methods include compost, incineration, and thermos drying, which are excellent for sludge treatment but present application limitations. Composting is restricted by the limited use of the final product, relatively large requirement for land, and unpleasant scent produced. Thus, the management of sludge also poses an economic and formidable environmental challenge for any effluent treatment facilities. Therefore, there is a need to identify cost-effective and sustainable solutions to the utilization of distillery sludge (Kumar and Sharma 2019; Chandra and Kumar 2017c). Thus, research is being carried out and attempts are being made to prepare this solid waste for proper use in the production of biofuels. The fat content of the distillery industry wastes is remarkable and there is almost no application method to recover these wastes. Given the lipid content of sludge, there is a promising potential in extracting these oils and converting them to a biofuel (e.g., biodiesel).

12.2 Distillery Waste

Distillery sludge is a solid waste that comes from the wastewater treatment plants during treatment of spent wash. It contains a variety of organic compounds, mainly lipids, proteins, sugars, and phenolics compounds. Moreover, distillery sludge also contains a wide range of harmful substances such as acetamide, 2,2,2-trifluoron-methyl-(TMS); 2-butanol, *tert*-butyldimethylsilyl ether; pyridine, 3-trimethylsiloxy; 3-hydroxy-6-methypyridine 1tms; ad 1,4-dimethylpyrrolo(1,2-*a*) pyrazine; 2-butanone, 4-[2-isopropyl-5-methyl]; 2,5-cyclohexadiene-1,4-dione, 2,6-bis(1,1-dimethylethyl); octadecanoic acid, methyl ester; 17-pentatriacontene; benzene, 1-ethyl-2-methyl, benzene, 1-ethyl-4-methyl benzoic acid, 3,4,5-tris(TMS oxy), TMS ester; hexanedioic acid, dioctyl ester; stigmasterol TMS ether; 5α-cholestane,4-methylene; campesterol TMS; and β-sitosterol and lanosterol and heavy metals (Chandra et al. 2018b; Kumar and Chandra 2020b). Despite the organic and inorganic pollutants, distillery sludge consists of a variety of microorganisms responsible for the biological treatment of spent wash (Chandra and Kumar 2017b). This population mostly consists of *Bacillus* and *Enterococcus* species that utilize the organic content of the wastewater for growth, either as part of their cellular structures or as energy and carbon storage compounds that are mostly lipidic in nature. Kumar and Chandra (2020b) have reported different groups of microorganisms from distillery sludge-contaminated sites. Bacterial members of the phylum Proteobacteria, Acidobacteria, Bacteroidetes Gematimonadetes, Verrucomicrobia, Firmicutes, Chloroflexi, Actinobacteria, Tenericutes, and Nitrospirae have been known to contribute directly or indirectly to distillery sludge's pollutants degradation. Rheinheimera, Sphingobacterium, and Idiomarina have been considered as "dominant taxa" as these organisms showed huge abundance in rhizospheric sludge associated with root of *Saccharum arundinaceum* grown on dumped organometallic sludge discharged from sugarcane-molasses-based distillery. Biodiesel production using third-generation feedstocks (microbial lipids from microalgae, yeast, and fungi) could be considered an effective alternative to replace edible oil feedstocks due to the high lipid yield and abundance of industrial waste to be used as a carbon source for microbial lipid production. Due to high lipid contents, distillery sludge can be used as a potential feedstock for biodiesel production.

12.3 Biodiesel and Their Production Processes

Distillery sludge, an abundant by-product (biomass) generated from biological wastewater treatment, which is the most widely used technology for wastewater treatment, is a major source of solid and groundwater pollution and must be adequately treated before disposal into the environment (Chandra and Kumar 2017c; Mahaly et al. 2018). One viable alternative to sludge management and disposal methods is to utilize the sludge as a source of lipid feedstock for biodiesel production. Several research has indicated that the lipids contained in industrial sludge are a potential feedstock for biodiesel production (Kumar and Thakur 2020a). Two different approaches have been investigated for the production of biodiesel from glycerides and free fatty acids (FFAs) extracted from sludge. The first one is a two-step process consisting of organic solvent extraction followed by acid-catalyzed transesterification of the isolated oil fraction. The second one is a one-step direct transformation consisting of the simultaneous extraction and conversion of the lipid fraction contained in sludge. Lipid is a natural mixture of triglycerides, diglycerides, monoglycerides, cholesterols, free fatty acids, phospholipids, sphingolipids, etc. Distillery sludge contains a significant amount of lipid fraction that is a composite organic matrix (characterized as oils, fats, and long-chain fatty acid) originating from the plant materials in the sludge, and/or from the phospholipids in the cell membranes of microorganisms, their metabolites, and by-products of cell lysis (Chandra and Kumar 2017b; Kumar and Chandra 2020b). Distillery sludge is considered a suitable substitute for raw biodiesel materials because of its two unique advantages: (i) it is widely and consistently available, and (ii) it contains considerable amounts of fatty acids such as tetradecanoic acid, 9,12-octadecadienoic acid, *cis*-10-nonadecenoic acid, hexadecanoic acid, hexacosanoic acid, decanoic acid, octadecanoic acid, and tetradecanoic acid (Chandra et al. 2018b; Kumar and Chandra 2020b). Besides, distillery sludge comprises various undesirable substances, such as recalcitrant organic pollutants, heavy metals, and indigenous microorganisms, which may lead to secondary pollution if not suitably dealt with thorough disposal and sludge treatment (Chandra and Kumar 2017b). In addition, sludge also contains abundant nutrients that are essential for the growth of microorganism. Studies have revealed that wastewater sludge could be used as a medium for the growth of microorganisms. The use of wastewater sludge as raw material to cultivate oleaginous microorganisms would reduce the cost of lipid production and mitigate the sludge disposal pressure. Wastewater sludge compositions vary according to the sludge type (primary, secondary, mixed) and wastewater source. Biodiesel is an alternative renewable, sustainable, biodegradable, oxygenated, sulfur-free, and less-toxic fuel and has a low environmental impact as compared to the petroleum-based diesel (Figure 12.1). It may be derived from a variety of feedstock such as vegetable oils, animal fats, used frying oils, microbial oils, or industrial sludge and it can be used in diesel cars without modification of the existing engine design. Apart from their renewable nature, biodiesel is a clean-burning fuel (for its oxygen content) with low exhaust emissions, free from sulfur, and carcinogen content.

Chemically, biodiesel, one of the first-generation biofuels, is produced in the form of a mixture of fatty acid alkyl ester from vegetable oils, microbial oil, or animal fats by esterification reaction of fatty acids (chain length C_{14}–C_{20}) with short-chain alcohols, primarily, methanol and ethanol, or transesterification reaction of triglyceride with short-chain alcohol in the presence of a catalyst (i.e., acid, base, and enzyme) that generate glycerol as a by-product (Figure 12.2). At industrial scale, production of biodiesel is carried

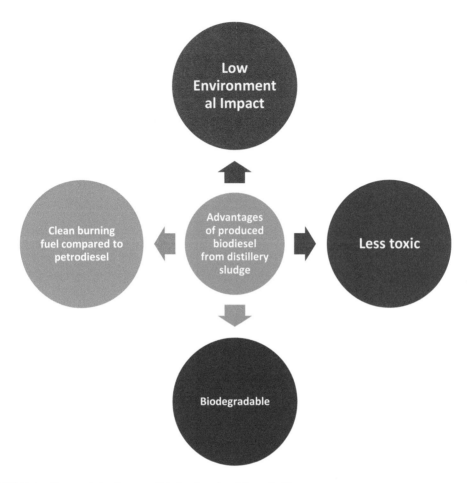

FIGURE 12.1 Characteristics features of biodiesel produced from distillery waste.

out via the transesterification process using triglycerides and methanol in the presence of homogeneous base catalysts (Kumar and Thakur 2020b). Biodiesel production is considered as one of the potential options among existing biofuels production technologies. Thus, biodiesel production has increased significantly in recent years, and the prices of biodiesel feedstocks have also risen substantially. However, a major constraint for the commercialization of biodiesel production using various feedstocks is the high cost of pure vegetable, animal, microbial, or seed oils, which constitutes between 70% and 85% of the overall biodiesel production cost. Therefore, the cost competitiveness of biodiesel has declined and

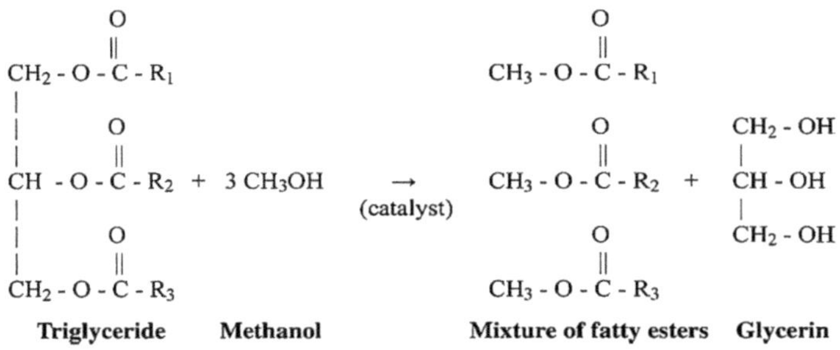

FIGURE 12.2 A typical transesterification reaction occurs between a triglyceride (from oils or fats) and a short-chain simple alcohols (usually methanol or ethanol) in the presence of a catalyst to generate the fatty acid methyl/ethyl esters.

biodiesel now struggles to compete with conventional fossil fuels with respect to price. Hence, in order to reduce the cost per gallon of biodiesel, alternative feedstocks that are readily available in large quantities and at a low cost must be used. In this context, high lipid/oil content of the feedstock such as distillery sludge can be a promising substrate for biodiesel production. Distillery waste is also considered an inexpensive and promising feedstock used for biodiesel production that can detract biodiesel production cost in the global market to make it more profitable for industries.

12.4 Transesterification Reaction and Their Advantages in Biodiesel Production

The reaction involves the breakage of the glycerol structure and exchanges of alkyl groups between the alcohol and ester part of the triglyceride molecule; as such, the reaction is known as transesterification. Biodiesel is commonly produced from pre-extracted oils by the process of transesterification, a reaction that occurs between an acylglycerol (from oils or fats) and a short-chain simple alcohols (usually methanol or ethanol) in the presence of a catalyst (i.e., bases, acids, or enzymes) to generate the fatty acid methyl/ethyl esters (Figure 12.2). In transesterification reaction, one mole of triglyceride reacts with three moles of short-chain alcohol (e.g., ethanol or methanol) in the presence of a suitable catalyst to produce biodiesel. In this process, glycerol is generated as a co-product of the reaction. Transesterification is regarded as the most suitable process for biodiesel production and so applied in the majority of industries. It is highly popular because of its low cost and simplicity and it reduces the viscosity of oil, making it suitable for engines and equipment. The transesterification reaction is used to break down the chemical structure of the triglycerides in oil via the exchange of the alkyl groups between an ester and an alcohol with the alcohol being used as a reactant.

The process of transesterification can be categorized as a catalytic and non-catalytic process. The catalytic process involves the use of a catalyst to enhance the reaction rate and course of the time. On the other hand, the non-catalytic process does not involve the use of any catalyst. On the basis of the type of catalysts used in the production process, the catalytic method can be further divided into homogeneous and heterogeneous catalysts. Homogeneous catalysts can be either acid or alkaline catalysts. Examples of acid catalysts include sulfuric acid, hydrochloric acid, ferric sulfate, phosphoric acid, etc. Acid catalysts are more suitable for the feedstocks having a high amount of free fatty acids and water contents as it is helpful in lowering the FFA content. Because transesterification process has difficulty converting into esters in the presence of FFA and water, it requires high-quality raw materials to avoid undesirable side reactions and hydrolysis (saponification) or additional pretreatment to remove the initial FFAs. On the other hand, heterogeneous catalysts include enzymes and other catalysts like titanium silicates, alkaline earth metal, amorphous zirconia, titanium, and potassium zirconia. Although homogeneous basic catalysts such as sodium hydroxide (NaOH) and potassium hydroxide (KOH), sodium methoxide, or sodium ethoxide are commonly used in the industrial production of biodiesel, there have several disadvantages such as corrosion problems, generation of huge volume of wastewater, and inability to reuse the catalysts. To address the issues created by the homogeneous catalysts, heterogeneous catalysts have the potential as they provide easier separation techniques, leave the product free of catalyst impurities, and exclude the requirement for product neutralization and purification steps. In addition, the lower consumption of these catalysts with high reusability could lead to economical production costs of biodiesel. Biodiesel can be produced from wastewater sludge via two methods: (i) ex situ transesterification and (ii) in situ transesterification. **Figure 12.3** describes the general process of biodiesel production by in situ and ex situ transesterification methods.

In ex situ transesterification methods, sludge is directly subjected to lipids extraction using organic solvents, and thereafter the extracted lipids are converted to biodiesel by reacting with methanol in the presence of suitable acid or base catalyst. Lipids are extracted from primary, secondary, blended, and stabilized sludge in a soxhlet extractor, using generally *n*-hexane as a solvent. Finally, the lipids are converted by catalytic transesterification into their corresponding FAME as a biodiesel. The extraction and transformation of the lipids could yield an unexploited source of cheap and readily available feedstock for biodiesel production. In order to produce FAME from microorganisms, there are several procedures that must be followed. First, microbial biomass must be cultivated in the stationary phase. Then, it must be dried

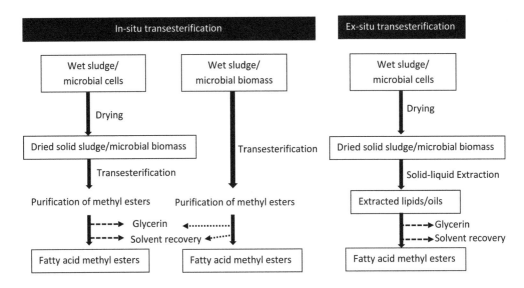

FIGURE 12.3 Simplified flow diagram of procedure for production of biodiesel by transesterification.

and disrupted mechanically or non-mechanically in order to release the intracellular lipids. Subsequently, the microbial lipids are recovered by liquid-liquid extraction using *n*-hexane or heptane, or either a mixture of hexane–ethanol or chloroform-methanol. Finally, the extracted lipids are esterified using methanol or ethanol as solvent and either strong acid (hydrochloric acid or sulfuric acid) or base (NaOH or KOH) as catalyst. Then, separation of the organic and inorganic phases of the mixture must occur, usually with *n*-hexane, to recover the esterified lipids. In direct acid- or base-catalyzed methanolysis, after the biomass recovery and drying steps, cell disruption, oil extraction, and esterification of the microbial oils occur simultaneously (one-step process), while drying as pretreatment step and FAME recovery using *n*-hexane are considered necessary. Although these acid or base catalysts are relatively cheap and highly sensitive to water molecules and FFAs in oils, the side reactions of hydrolysis and saponification, respectively, can occur. On the other hand, in situ transesterification method (also called direct transesterification) involves mixing of dried sludge with an organic solvent and acid catalyst to convert lipids to biodiesel, followed by separation of FAME from the organic solvent and thereafter the FAME is subjected to evaporation to recover solvents from FAME. This direct transesterification process removes additional lipid extraction and fuel conversion steps. This direct in situ transesterification process removes additional steps (i.e., the need to extract and separate the lipids and fatty acids contained in the sludge) by combining these two steps into a single step that reduce the amount of solvent, maximize the FAME yield, and overall reduce the cost of biodiesel production. Moreover, in the case of using wet cells, the spent time and energy for the drying step could also be reduced. Biodiesel production from primary and secondary sludges with and without previous extraction of lipids (in situ reactions) using sulfuric acid as a catalyst has been reported by several researchers. However, the yield of biodiesel production depends on several factors, including the type of catalyst, solvent to lipid ratio, FFAs and water content, as well as temperature and reaction time. The optimum production of biodiesel is faced with huge challenges. First, the lipids containing FFAs are usually extracted and then transesterified. Impurities in the lipids from the sludge would have interfered with the catalytic process in the conventional production of biodiesel.

12.5 Integrated Treatment of Distillery Effluent and Biodiesel Production

Many researchers have investigated lipid production through oleaginous microbes such as moulds, yeasts, and algae exhibiting the capacity to accumulate intracellular lipids in excess of 20–25% of their biomass during metabolic stress periods, especially during the stationary growth phase. Therefore, microbial lipids have the potential to be used as a raw material in biodiesel production. Combining wastewater treatment and biofuel production is considered a cost-effective way for better waste remediation and lowering the environmental impact for biofuel production. Phytofiltration systems are eco-friendly technologies for wastewater treatment, and they simultaneously serve as a source of biomass for biofuel production. The color removal from anaerobically digested sugarcane stillage by biomass from invasive macrophytes as phytofilters has been reported by Sánchez-Galván et al. (2015). In this study, the ability of untreated and acid-treated biomass from *Pistia stratiotes* (PL and APL, respectively) and *Eichhornia crassipes* (ELS and AELS, respectively) to remove color from anaerobically digested sugarcane stillage was investigated by authors. Decolorization of anaerobically digested sugarcane stillage also increased when the untreated biomass concentration was higher (15.41 ± 0.3 and $27.89 \pm 0.2\%$ for 5 and 10 g/L, respectively, for ELS; 15.61 ± 0.11 and $33.06 \pm 1.09\%$ for 5 and 10 g/L, respectively, for PL). The highest rate of color removal obtained using acid-treated biomass was 55.58 ± 1.82 and $56 \pm 0.77\%$ for AELS and APL, respectively. Similarly, the use of dry biomass from invasive macrophytes is an effective tool for color removal from anaerobically digested sugarcane ethanol stillage. Sánchez-Galván and Bolaños-Santiago (2018) investigated the phytofiltration of anaerobically digested sugarcane stillage using *Azolla* sp. The nutrient concentrations especially NH_4-N, PO_4-P, and SO_4-S in stillage were reduced significantly. The results of this findings indicated that a phytofiltration system involving *Azolla* sp. is efficient for reducing the organic matter content, nutrients, and color intensity significantly in ADS under the tested conditions. The conversion of nutrients from stillage into valuable *Azolla* biomass may provide an effective way to produce a very attractive feedstock for the production of a wide spectrum of biofuels. Further work at a larger scale is recommended by authors to confirm the feasibility of the laboratory-scale research findings; additionally, a cost analysis should be performed. Chuppa-Tostain et al. (2018) introduces and analyzes a novel system for production of biodiesel from *Aspergillus niger* biomass on sugarcane distillery wastewater. The soft-rot fungus *A. niger* demonstrated a great ability to grow on vinasse and to degrade this complex and hostile medium. The high biomass production is accompanied by a utilization of carbon sources like residual carbohydrates, organic acids, and more complex molecules such as melanoidins. Authors also showed that intracellular lipids from fungal biomass can efficiently be exploited into biodiesel. Anbarasan et al. (2018) demonstrated the production of biodiesel from lipids accumulated by *Metschnikowia pulcherrima* grown in distillery wastewater under varying conditions of pH, temperature, culture times, etc. The raw wastewater had a COD of 86 g/L and total dissolved solids of 46.9 g/L. The culture conditions for maximum growth were found to be pH 6.2, 30 °C, and 120 h. The lipid extraction was done and lipids were used for biodiesel conversion. In situ transesterification reaction was effected by base catalysis using NaOH and methanol to form fatty acid methyl esters. The yield reached up to 1.4 g/L. Malik et al. (2019) showed nanocatalytic ozonation of biomethanated distillery wastewater for biodegradability enhancement, color, and toxicity reduction with biofuel production.

12.6 Summary

i. Biodiesel is a promising, renewable, non-toxic, biodegradable, energy resource and has burning clean characteristics produced from various lipid sources.

ii. Biodiesel production has grown in popularity due to its environmental benefits and the need of an exact substitute for conventional diesel.

iii. It can be used directly in conventional engines. Compared to fossil fuels, it has advantages of being free of sulfur, featuring an inherent lubricity with higher cetane and flash point values.

iv. Biodiesel is not competitive with petro-diesel in terms of the production cost. The major cost (70% of total) of biodiesel production is from the utilization of feedstock oils and fats.

v. At present, biodiesel is more expensive than petro-diesel because it is mostly made from expensive virgin vegetable oils.

vi. Biological wastewater treatment plant facilities produce plenty of primary and secondary sludge, which is a potential non-food feedstock that offers significant potential for biodiesel production.

vii. The pollution caused by the distillery industry wastes may be reduced and more valuable products can be obtained by converting them to biodiesel.

viii. Biodiesel produced from activated sludge provides an opportunity to improve low-cost biofuel.

ix. When sludge is used as a lipid source, it was expected that the cost of biodiesel production would be highly reduced as sludge was cost-free material. In addition, wastewater sludge has been found as a suitable medium to cultivate microorganism due to the fact that the sludge was rich in carbon, nitrogen, and phosphorus.

x. Chemically, biodiesel is a fatty acid alkyl ester, commonly known as fatty acid methyl ester, that is produced via esterification and/or transesterification of lipid sources with alcohols in the presence of a base, acid, enzyme, or solid catalyst.

xi. The process of transesterification removes glycerin and the triglycerides replaces it with the alcohol used for the conversion process. This process decreases the viscosity but maintains the cetane number and the heating value.

xii. FAME can be used as an alternative diesel fuel as pure biodiesel or biodiesel blend.

xiii. Biodiesel is generally produced by transesterification of feedstock oil with alcohols in the presence of an acid/base catalyst.

13

Rules, Policies, and Laws Made by Government of India for Recycling, Reuse, and Safe Disposal of Distillery Waste into the Environment

13.1 Introduction

In India, bioethanol is mainly produced by the fermentation of diluted sugarcane molasses, a by-product generated during the sugar production process from sugarcane. After fermentation, alcohol is separated by distillation and the residual dark color liquid is discharged as vinasses/molasses wastewater/distillery spent wash, or effluent having a high organic and inorganic load. This effluent is dark brown in color, of low pH, and contains all the ingredients found in molasses except fermentable sugar. For every liter of ethanol produced, about 12–15 L of effluent are generated. The generated dark brown color effluent poses a serious pollution threat to the environment and human health. In an aquatic ecosystem, distillery effluent reduced penetration of sunlight in lagoons, lakes, and rivers, which in turn decreases both dissolved oxygen and photosynthetic activity, thereby causing aquatic life to suffer and resulting in deterioration of water quality and loss of productivity to such an extent that the water becomes unusable. As per the Ministry of Environment, Forests, and Climate Change (MoEF&CC), Government of India (GoI), alcohol distilleries and fermentation industries are listed at the top of "Red Category" industries having a high polluting potential; thus, distilleries must take appropriate measures to comply with the discharge standards set by the Central Pollution Control Board of India (CPCB), which is the national agency of GoI responsible for environmental compliance. Besides, new discharge regulation on fermentation industries (including alcohol distilleries) has limited the biochemical oxygen demand (COD), biological oxygen demand (BOD), and suspended solids (SS) in the final effluents as low as 150, 50, and 50 mg/L, respectively. Therefore, there is an emergent need to treat such high-strength distillery wastewater more efficiently and sustainably. In 2003, the CPCB stipulated that distilleries should achieve zero discharge in inland surface watercourses by the end of 2005. Moreover, Bureau of Indian Standard (BIS) provides guidelines to the authorities of state and central government that would help them to decide on suitable restrictions on liquid effluent disposal and the industry for selecting appropriate technology, suitable site, and the degree of treatment required for liquid effluents before their disposal. The CPCB has laid down rather rigid standards for the treatment and disposal of distillery effluent, which are difficult to achieve in India. Thus, the CPCB under the direction of GoI has modified the environment protection rules and allowed utilization of pretreated effluent for the ferti-irrigation of land. In addition, distilleries used several conventional treatment technologies to reduce the degradable organic matter to a respectable limit before discharging the effluent into the streams or on land. This chapter provides an overview of the various rules, policies, and laws made by GoI for recycling, reuse, and safe disposal and management of distillery waste in the environment.

13.2 Distillery Listed in "Red Category"

On March 5, 2016, Minister of State (Independent Charge) of MoEF&CC released the four-color categorization of industries as per the Pollution Index (PI) score between 0 and 100, as shown in **Table 13.1** (https://www.mapsofindia.com/my-india/government/colour-codes-of-industries-for-environment-clearances).

DOI: 10.1201/9781003029885-13

TABLE 13.1

The Four-Color Categorization of Industries as Per Their Pollution Index

Category	Characterization
Red Category	Industrial sector having Pollution Index score of 60 and above is called Red Category of industry. Normally, industries falling under "Red" Category will not be permitted in ecologically fragile or sensitive areas. Interestingly, industries like integrated automobile manufacturing, airports and commercial air strips, and milk and dairy products are included in the "Red" Category. There are 64 types of polluting industries that are classified as "Red Category" industries based on their emissions/discharges of high/significant polluting potential or generating hazardous wastes. Distillery industries, including fermentation industry, are identified by MoEF&CC as heavily polluting and covered under central action plan
Orange Category	Industries with PI score of 41–59
Green Category	Industries with PI score of 21–40
White Category	Industries with PI score including and up to 20 are categorized as "White" Category. This category covers 36 industrial sectors that are mostly non-polluting. Industries falling under "White" category include LED and CFL bulb assembly, power generation using solar photovoltaic technology, wind power generating units, hydel units less than 25 MW, products made from rolled PVC sheets using automatic vacuum forming machines, cotton and woolen hosiers using dry processes, etc.

PI: Pollution Index; LED: light-emitting diode; CFL: compact fluorescent lamp; PVC: polyvinyl chloride.

Based on the pollution levels, a PI score is assigned. PI is a function of emission (air pollutants), effluent (water pollutants), hazardous wastes generated, and consumption of resources. The PI of any industrial sector is 0–100 and the increasing value of PI denotes the increasing degree of pollution load from the industrial sector. The PI was derived after extensive consultation with various pollution control and monitoring bodies like CPCB, state pollution control boards (SPCBs), and MoEF&CC.

According to this list released on March 5, 2016, by the Minister of State (Independent Charge) of MOEF&CC, GoI, distilleries have been categorized under the category of "Red Category" and listed at the 65 positions in the list of highly polluting industries. Thus, it has become necessary to reduce freshwater consumption in distilleries, which in turn may reduce the generation of spent wash.

13.3 Development of Environmental Standards and Regulation for Indian Distilleries

Based on the brainstorming session among CPCB, SPCBs, and MoEF&CC, the MoEF&CC formulates and notifies standards for emission or discharge of environmental pollutants, namely, air pollutants, water pollutants, and noise limits, from industries, operations, or processes with an aim to protect and improve the quality of the environment and abate environmental pollution. The standards for any industrial process or operation recommended by CPCB are subjected to stakeholder consultation, including the general public. The comments are compiled and techno-economically examined by CPCB and change, if any, incorporated. The modified standards are placed before the "Expert Committee" (EC) of MoEF&CC for approval. The EC of MoEF&CC comprises representatives from industry associations, subject experts, and concerned ministries of the industrial sectors, besides the officials of MoEF&CC and CPCB. The EC recommended standards for approval and legal vetting are published in Gazette of India, a public journal and an authorized legal document of the GoI, published weekly by the Department of Publication, Ministry of Housing and Urban Affairs. As a public journal, the Gazette prints official notices from the government. It is authentic in content, accurate, and strictly in accordance with the government policies and decisions. During the year, Standards with respect to the following category of industries have been notified at http://moef.gov.in/environment/pollution/. The notification of standards also involves the formulation of load-based standards, i.e., discharge limits of pollutants per unit of product obtained/processes performed, to encourage resource utilization efficiency and conservation aspects.

13.4 Environmental (Protection) Acts

In India, the Environmental (Protection) Act was enacted in 1986 with the objective of providing for the protection and improvement of the environment. It empowers the Federal Government to establish authorities (under Section 3(3)) charged with the mandate of preventing environmental pollution in all its forms and to tackle specific environmental problems that are peculiar to different parts of the country. The Act was last amended in 1991. The legal documents of Environmental (Protection) Act, 1986, is available on the official website of CPCB as shown in Figure 13.1.

Under the Environmental (Protection) Act, the GoI has specified minimum national standards (MINAS) for different industries taking into account the characteristics of the effluent and the minimum acceptable quality of the treated effluent. The standards for the distillery industry stipulate that the treated effluent should have a pH in the range of 5.5–9.0, maximum BOD level of 30 mg/L for disposal into inland surface waters, and 100 mg/L for disposal on land. It also states that all efforts should be made to remove color and unpleasant odor as far as practicable. The general standard of effluent discharge from Indian industries are presented in Tables 13.2 and 13.3.

13.5 The Central Pollution Control Board

The CPCB has been playing a vital role in abatement and control of pollution in the country by generating environmental quality data, providing scientific information, formulating national policies and programs, and training and promoting awareness. It has been actively involved in developing the sectorwise standards at national level for effluents and emissions from different polluting industrial sectors and formulating nationwide programs for their effective implementation. The CPCB in pursuance of its mandate formulates national programs for prevention and control of pollution in India. These include nationwide monitoring network, laying down national standards for ambient water and air quality, source-specific MINAS for effluents and emissions, and action plans for critically polluted areas and highly polluting categories of industries. Concerning polluting industries, the state pollution control boards (SPCBs) in state and the pollution control committees (PCCs) in Union Territory Administration enforce the standards laid down for various types of industry. For a nationwide drive to control industrial pollution, the CPCB enlisted 17 categories of highly polluting industries and grossly polluting industries discharging their effluents into the rivers and lakes. The SPCBs/PCCs were asked to give special attention to these industries and the progress of implementation of the program is regularly monitored by CPCB. Earlier, such an initiative was also taken with regard to the grossly polluting industries discharging their effluents into the river Ganga. The CPCB performs functions as laid down under The Water (Prevention & Control of Pollution) Act, 1974, and The Air (Prevention and Control of Pollution) Act, 1981. In 2003, CPCB constituted a task force on Corporate Responsibility for Environmental Protection (CREP) that stipulated that distilleries

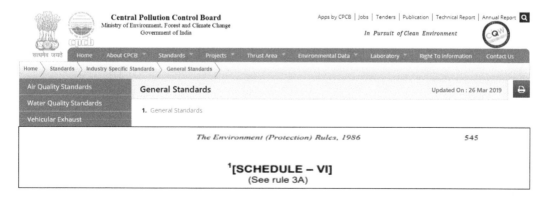

FIGURE 13.1 General view of a webpage of Central Pollution Control Board.

TABLE 13.2

General Standard for Discharge of Environmental Pollutants Part A: Effluent

S. No.	Parameter	Inland Surface Water	Public Sewers	Land for Irrigation	Marine Coastal Areas
			Standards		
1.	Color and odor	See 6 of Annexure I	–	See 6 of Annexure I	See 6 of Annexure I
2.	Suspended solids (mg/L) =	100	600	200	For process wastewater: 100; for cooling water effluent: 10% above total suspended matter of influent
3.	The particular size of suspended solids	Shall pass 850 μm IS sieve	–	–	Floatable solids, max 3 mm; settleable solids, max 850 μm
4.	pH value	5.5–9.0	5.5–9.0	5.5–9.0	5.5–9.0
5.	Temperature	Shall not exceed 50 °C above the receiving water temperature	–	–	Shall not exceed 50 °C above the receiving water temperature
6.	Total residual chlorine (mg/L)	1.0	–	–	1.0
7.	Ammonical nitrogen (as N) (mg/L)	50	50	–	50
8.	Total Kjeldahl nitrogen (as NH_3) (mg/L)	100	–	–	100
9.	Free ammonia (as NH_3) (mg/L)	5.0	–	–	5.0
10.	Biochemical oxygen demand (3 days at 27 °C) (mg/L)	30	350	100	100
11.	Chemical oxygen demand (mg/L)	250	–	–	250
12.	Arsenic (As) (mg/L)	0.2	0.2	0.2	0.2
13.	Mercury (Hg) (mg/L)	0.01	0.01	–	0.01
14.	Lead (Pb)	0.1	1.0	–	2.0
15.	Cadmium (Cd)	2.0	1.0	–	2.0
16.	Hexavalent chromium (Cr^{+6})	0.1	2.0	–	1.0
17.	Total chromium (Cr)	2.0	2.0	–	2.0
18.	Copper (Cu)	3.0	3.0	–	3.0
19.	Zinc (Zn)	5.0	15	–	15
20.	Selenium (Se)	0.05	0.05	–	0.05
21.	Nickel (Ni)	3.0	3.0	–	5.0
22.	Cyanide (CN)	0.2	2.0	0.2	0.2
23.	Fluoride (F)	2.0	15	–	15
24.	Dissolved phosphate (as P) mg/L	5.0	–	–	–
25.	Sulfide (S)	2.0	–	–	5.0
26.	Phenolic compounds (as C_6H_5OH) (mg/L)	1.0	5.0	–	5.0

TABLE 13.2   *(Continued)*

S. No.	Parameter	Inland Surface Water	Public Sewers	Land for Irrigation	Marine Coastal Areas
			Standards		
27.	Radioactive materials				
(a)	Alpha emitter micro curie/mL	10^{-7}	10^{-7}	10^{-8}	10^{-7}
(b)	Beta-emitter micro curie/mL	10^{-6}	10^{-6}	10^{-7}	10^{-6}
28.	Bioassay test	90% survival of fish after 96 h in 100% effluent	90% survival of fish after 96 h in 100% effluent	90% survival of fish after 96 h in 100% effluent	90% survival of fish after 96 h in 100% effluent
29.	Manganese (Mn)	2 mg/L	2 mg/L	–	2 mg/L
30.	Iron (Fe)	3 mg/L	3 mg/L	–	3 mg/L
31.	Vanadium (V)	0.2 mg/L	0.2 mg/L	–	0.2 mg/L
32.	Nitrate nitrogen	10 mg/L	–		20 mg/L

TABLE 13.3

Wastewater Generation Standards Part B

S. No.	Fermentation Industries	Quantum
1.	Maltry	3.5 m^3/ton of grain produced
2.	Brewery	0.25 m^3/kL of beer produced
3.	Distillery	12 m^3/kL of alcohol produced

should achieve zero discharge in inland surface watercourses by end of 2005. CPCB guidelines for the treatment and disposal of spent wash from molasses-based distilleries are available at their official web-page (https://cpcb.nic.in) (Figure 13.2).

According to the Environment (Protection) Rules 1986, CPCB has prescribed safe disposal limits for distilleries (**Table 13.4**).

13.6 Scheme of Common Effluent Treatment Plants

The concept of common effluent treatment plants (CETPs) was envisaged to treat the effluent emanating from the clusters of compatible small-scale industries. It was also envisaged that the burden of various government authorities working for controlling pollution and monitoring of water pollution could be reduced once the CETPs are implemented and commissioned. The main objective of the CETPs is to reduce the treatment cost to be borne by an individual member unit to a minimum while protecting the environment to a maximum. Effluent treatment and water conservation are the prime objectives of the CETP. A centrally sponsored scheme (CSS) had been undertaken by the government for enabling SSI to set up new and upgrade the existing CETPs to cover all the states in the country. The CSS of CETPs had been revised by the ministry since 2012 with the following salient features:

   i. The Central subsidy has been enhanced from 25% to 50% of the project cost.

  ii. All the three levels of treatment, primary, secondary, and tertiary, are to be covered for assistance. Progressive technologies like zero liquid discharge will also be considered for assistance, subject to a ceiling.

 iii. The management of the CETP is to be entrusted to a Special Purpose Vehicle registered under an appropriate statute.

 iv. Performance guarantee at full design load is to be ensured upfront.

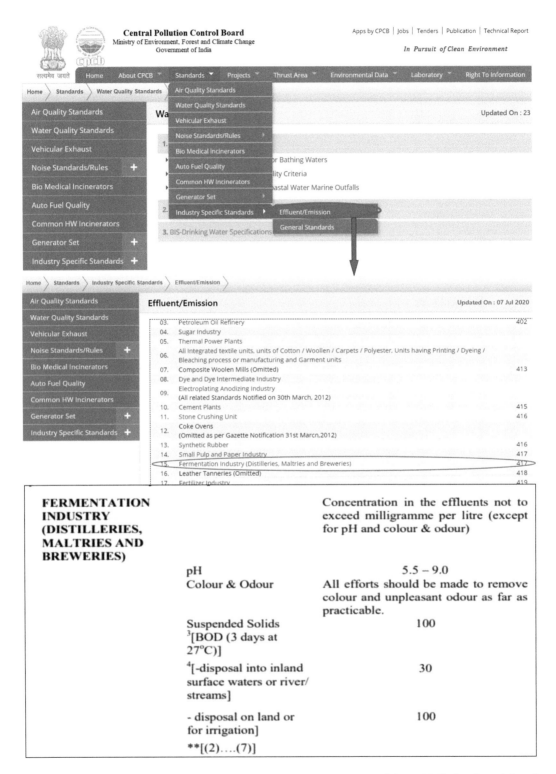

FIGURE 13.2   The industrial effluent discharges norms listed on a webpage of Central Pollution Control Board (CPCB).

TABLE 13.4

Safety Disposal Limits Prescribed by CPCB for Distilleries, Environment (Protection) Rules, 1986

S. No.	Parameter	Standard
1.	pH	5.5–9.0
2.	Color and odor	All efforts should be made to remove color and unpleasant odor as far as practicable
3.	Suspended solids	100 mg/L
	BOD (3 days at 27 °C)	30 mg/L
	(Disposal into inland surface water or river/streams)	100 mg/L
	Disposal on land or for irrigation)	

Concentration in the effluent not to exceed mg/L (except for pH, color, and odor).

However, after the evaluation of the Plan Scheme of MoEF&CC in 2016–2017, it was decided to discontinue the CETP scheme after funding support to the existing ongoing projects.

13.7 Corporate Responsibility for Environmental Protection

For every distillery in the country, it is now mandatory to achieve "zero spent wash discharge" as per CPCB. Since the standards prescribed are very stringent and difficult to practically achieve, a Charter on CREP (CPCB 2003) was proposed for distilleries. The MoEF&CC launched the Charter on CREP in March 2003 with the purpose to go beyond the compliance of regulatory norms for prevention and control of pollution through various measures, including waste minimization, in-plant process control, and adoption of clean technologies (http://www.cpcb.nic.in/ar2003/ar2-3ch10.htm.). The Charter has set targets concerning conservation of water, energy, recovery of chemicals, reduction in pollution, elimination of toxic pollutants, and management of residues that are required to be disposed of in an environmentally sound manner. The Charter enlists the action points for pollution control for various categories of highly polluting industries. Furthermore, in 2003, CPCB constituted a task force on CREP that stipulated that distilleries should achieve zero discharge in inland surface watercourses by end of 2005 (CPCB, 2003). This was a voluntary and consultative initiative of the MoEF&CC that has recommended the following technologies/processes for spent wash treatment:

- Reboiler
- Biomethanation
- Reverse osmosis (RO) system
- Multi-effect evaporator (MEE)
- Biocomposting and one time controlled land application
- Ferti-irrigation
- Turbo mist evaporation
- Concentration and incineration

Effective handling and disposal of the generated effluent is a major concern in all Indian distilleries since the units are required to meet the discharge standards (**Table 13.4**) laid down by CPCB. The industry was advised to treat its effluent to achieve zero discharge and no liquid discharge was allowed either on land or in any water body. To comply with the disposal norms, the CPCB recommends one or a combination of the following schemes to treat the generated wastewater (CPCB, 2003). After thorough study of the processes and technologies, it was left to the industry to adopt any one or combination of two to three technologies out of the above-mentioned various technologies.

13.8 CREP Conditions for Existing Molasses-Based Distilleries

Existing molasses-based distilleries to achieve zero effluent discharge by December 2005, through any or combination of the following measures:

1. Compost making with press mud/agricultural residue/municipal waste. Concentration and drying/incineration.

2. Treatment of spent wash through biomethanation followed by two-stage secondary treatment and dilution of the treated effluent with process water for irrigation as per norms prescribed by CPCB/MoEF.

3. Treatment of spent wash through biomethanation followed by secondary treatment (BOD < 2500 mg/L) for controlled discharge into sea through a proper submerged marine outfall at a point permitted by SPCB/CPCB in consultation with National Institute of Oceanography (NIO), so that dissolved oxygen in the mixing zone does not deplete less than 4.0 mg/L.

4. For taking a decision on the feasibility of one-time controlled land application of treated effluent, a study will be undertaken within three months.

5. Till 100% utilization of spent wash is achieved, controlled and restricted discharge of treated effluent from lined lagoons during rainy season will be allowed by SPCB/CPCB in such a way that the perceptible coloring of river water bodies does not occur.

13.9 Consent Conditions According to CREP Norms for Distilleries

1. Storage capacity of raw spent wash, treated effluent, shall not exceed 30 days. The lagoons should be lined with HDPE sheets and protected with stone/concrete/brick lining.

2. The compost yard shall be lined with HDPE sheets and protected with brick/concrete/stone lining. Provisions should be made for leachate collection gutter and sump well.

13.10 Modifications in the Recommendations of CPCB for Distilleries (2008)

Subsequent to the problems associated with distilleries due to the treatment methods used, CPCB modified the recommendations for composting, ferti-irrigation, and one-time land application of spent wash. The modified recommendations are as follows:

1. Ferti-irrigation and one-time land application of spent wash by new stand-alone distilleries may not be considered by SPCB/MoEF/PCC.

2. Establishment of new distilleries attached with sugar units may be considered if one of the following options is followed; biomethanation followed by biocomposting; or reboiler/evaporation/concentration followed by incineration of concentrated spent wash in boiler (for power generation).

3. Existing stand-alone distilleries may not consider an increase of production/expansion based on composting, ferti-irrigation, and one-time application of spent wash.

4. Existing distilleries (both stand-alone and those attached with sugar units) that does not comply with the required environmental standards may be asked to switch over to emerging technologies from existing technologies of composting, ferti-irrigation, and one-time land application of spent wash in a time-bound manner.

13.11 Observations of the Expert Subgroup

The Expert group visited distilleries representing stand-alone distilleries, distilleries attached with sugar factories, and most modern distillery producing rectified spirit, alcohol, and fuel ethanol. The observations of the Subgroup are briefly presented below:

i. It was observed by the committee that popularly in the industry circle the distilleries are viewed as the effluent treatment plants of the sugar industry. If there were no molasses-based distilleries, the disposal of the entire molasses as effluent would be the responsibility of the sugar industry.

ii. In molasses, the presence of potassium, sodium, etc. is carried from the cane itself but the increased levels of calcium, phosphorus, and sulfur are due to the use of "milk of lime" during the clarification process and sulfitation process. Alternative methods of clarification in place of "milk of lime" may improve the quality of molasses by reducing these elements, rendering the usability of biomethanated spent wash for agriculture applications.

iii. The effluent discharge guidelines for distilleries having biocomposting system are not allowed to work during the rainy season. Therefore, distilleries can operate only for a maximum of 270 days a year.

iv. Most of the distilleries have biomethanation plant and biocomposting systems and few distilleries have installed RO plant to reduce the effluent quantity and very few have biogas power plants. The subgroup was informed about the operational difficulties of RO system. It frequently gets clogged necessitating frequent maintenance and quick change of membranes. This is resulting in high operational cost of RO system.

v. The modern distillery has adopted various available technologies and achieved improvements with respect to productivity, quality, proper management of effluent treatment, wastewater recycle for process and non-process applications, minimum freshwater consumption, etc. This distillery had also installed a new multiple effect evaporation (MEE) plant for concentration of biomethanated spent wash and Process Condensate Treatment Plant (PCTP) for stripping of NH_3 and further treatment of process condensate. These technologies have helped the unit to recycle and reduce its freshwater consumption. The freshwater consumption for this unit, as stated, is 6.40 kL/kL of alcohol production after considering recycle streams. The unit has achieved minimum freshwater consumption, minimum concentrated spent wash generation, and a maximum ratio of spent wash to press mud cake in a biocomposting process that has resulted in near-zero discharge of spent wash.

vi. The distilleries have adopted selectively some of the improved technologies as appropriate. These are reboilers along with distillation column, integrated evaporation system to concentrate the spent wash, aerobic biocomposting, membrane filtration, RO with different degree of success, multiple-effect evaporation, aerobic treatment (secondary or tertiary treatment), ferti-irrigation, controlled land application, turbo-mist evaporation, incineration of concentrated spent wash, gasification of distillery spent wash, co-processing in cement, steel, and power plants.

vii. The spent wash is generated on an average 10–12 L per liter of alcohol produced and is treated first in the biomethanation plant to reduce the BOD by 85–90% and COD by 60–65%. The biomethanated spent wash is then passed through RO and/or MEE reducing the quantum of waste by 45–50% and the permeate/process condensate is recycled. The rejects/concentrates are used in biocomposting. The biocomposting plant has modern facilities for windrow preparation, windrow dressing, spent wash application, aeration and turning, etc. The biocompost prepared is further enriched with biofertilizers and micronutrients.

viii. During the visit to one of the mills in the southern region, it was learnt that out of the 1000 m³ of biomethanated spent wash generated every day by the distillery, 600 m³ is consumed in the tertiary biocomposting system. The balance is used in controlled land application under the supervision of an agricultural university as per the protocol of CPCB. The plant applies spent wash to about 10 acres of land every day. About 4000 acres of agricultural and is required every year for controlled land application (CLA). The management of the distillery informed that there is consistent demand for biomethanated spent wash from nearby farmers because of the observed improvement in crop yield. It was also informed that there are no reports of groundwater pollution and any harmful effect on soil fertility. During the visit, the committee also interacted with the farmers and visited the agricultural field where this spent wash is being applied.

ix. Availability of press mud for mixing with spent wash for biocomposting is a major problem for stand-alone distilleries.

x. As per CREP norms, only 30 days spent wash storage is allowed. The storage tank is also required to be made impervious as per CREP norms by putting 250 μm HDPE sheet. The industry felt that it is too short a period and requested that a storage of 90 days be allowed so that distilleries can operate throughout the year to minimize the production losses.

xi. As per CREP norms, it is mandatory for all distilleries in the country to achieve zero spent wash discharge. In practice, it is very difficult to achieve zero spent wash discharge.

The CPCB had prescribed Guidelines for Water Quality Management in January 2008. These guidelines have recognized that the reuse and recycling of wastes for the agricultural purposes would not only help to reduce the pollution and requirements of freshwater for such use but would also supplement the much needed nutrients and organic manure to plants. These guidelines also state that the resolution under CREP should also be adhered to. The rich nutrient contents of spent wash make its use in agriculture very viable. Pre-sown land application, biocomposting, and ferti-irrigation have been variously applied. These technologies had been accepted for implementation through the Charter for CREP decided between the CPCB and distillers. The CPCB is, however, through a resolution of May 2008, not encouraging stand-alone distilleries to practice these technologies. However, distilleries attached to sugar units may follow biomethanation and biocomposting and concentration and incineration technologies. Crop or soil application is not being encouraged. It has also been suggested that attached distilleries that are not achieving the standards should change from composting, one-time land application, and ferti-irrigation in a time-bound manner. It only says that ferti-irrigation, composting, and land application options did not work in rainy seasons. Interestingly, the said minutes, which discourage the use of spent wash in ferti-irrigation, composting, and land application, on the very next page, also recognize that "As per CPCB, the bio-compost contains 2.5% nitrogen, 1.8% phosphorus and 3% potassium.

13.12 Need to Use a Mix of Options

There is a need to use a mix of treatment options with adequate protocols and guidelines so that spent wash can be gainfully utilized for biogas generation, biocompost, ferti-irrigation, one-time land application, sodic land reclamation, and co-processing. Unit-specific combinations could be prescribed based on the availability of land, sodic land, clinker manufacturing units and other furnaces, etc. The guidelines given by the CPCB for co-processing of concentrated spent wash in cement kilns require certain modifications in the cement kilns. The transportation of spent wash from the distillery to the cement factory also require careful handling in tankers specially designed for this purpose. Air pollution concerns are also to be addressed. The co-processing plants have to be designed, equipped, built, and operated in such a way as to prevent emission into the air giving rise to significant ground-level air pollution; in particular, exhaust gases have to be discharged in a controlled fashion by means of a stack, the height of which is calculated in such a way as to safeguard human health and environment. The co-incineration initiative of the CPCB suggested burning concentrated spent wash as fuel in cement/steel industries along with other fuels/raw materials. Co-processing of spent wash concentrate has been recommended to the extent of 3.0–3.5% of heat/coal substitution—a concentration up to which co-processing of the spent wash has not been considered to influence clinker quality or change kiln behavior. A maximum loading of 1000 L/h has been experimented. This would only mean about 24 KLD of concentrated or about 50 KLD of raw spent wash per day in a 3000 TPD cement kiln. This is a very small quantity as compared to the total spent wash available. Also the availability of willing cement mills, furnaces, and TPPs may be limited in the area of the distillery. Uttar Pradesh has been reported to have 56 molasses-based distilleries, with only 1 cement plant and 11 TPPs. This would again support the view of a mix of technologies with biocomposting, ferti-irrigation, and one-time land application also being considered on account of their ecological and agronomical advantage. Even though the initial trials on co-processing appear to be encouraging, the effect of inorganic constituents in the spent wash on the finished product is to be assessed and the applicability of the technology for distilleries that are located far away is to be assessed in terms of cost-effectiveness.

14

Emerging Issues, Challenges, and Future Outlook of Distillery Waste Management

Molasses-based distilleries are among the most polluting industries generating a large bulk volume of high-strength effluent. Decolorization of distillery effluent (DE) is a critical concern and an understanding of the structure and characteristics of the color-causing components is required to develop an appropriate technique for their removal. Although many physicochemical and biological technologies are developed to treat the color effluent, none of the processes showed convincing results to achieve safe and economically feasible disposal due to the energy-intensive, high cost of chemicals as well as the associated sludge handling. Hence, the treatment of color liquid effluent generated from alcohol distillery industries is a big challenge for modern technologies combining high effectiveness of degradation of pollutants with low costs of the process across the globe. The major challenges of distillery effluent are the removal of total dissolved solids and degradation of color-contributing compounds like melanoidin, polyphenols, caramel, and alkaline degradation products of hexose and their metabolic products as well as inorganic compounds before its safe disposal into the environment. Moreover, distilleries are facing the following challenges during decolorization and detoxification of distillery effluent.

i. In most of the decolorization and degradation studies, melanoidins have been considered to be the main coloring pigments. However, there are several colorants instead of one type of pigment (melanoidin) alone that may result in the undesirable color, which enhances the difficulties of efficient decolorization and degradation of DE. Thus, advanced treatment techniques are required to degrade compounds that contribute to color from DE.

ii. Melanoidins and polyphenols exhibit antioxidant and antimicrobial properties that make difficult the treatment of DE at the industrial level. Thus, they inhibit the growth of potential microorganisms used in aerobic and anaerobic biological treatment processes, thus slowing down the treatment process of DE.

iii. Melanoidins show variable absorption range in the ultraviolet and visible region that makes more difficult to understand the mechanism of melanoidins degradation and decolorization and further characterization of its metabolic products.

iv. The maximum absorbance (λ_{max}) of melanoidins discharges in DE has been recorded at a wavelength of 475 nm by UV-visible spectrophotometer, but this value is not steady; it sometimes varied depending on the interaction of melanoidins with co-pollutants. The changeable λ_{max}, thus, complicates the understanding of the mechanism of decolorization and degradation of melanoidin and characterization of its fragmentary products.

v. It has been widely reported that due to the anionic nature of melanoidins in DE, it binds with a mixture of various trace elements and some other organic compounds that makes more difficult its biological degradation.

vi. Different amino acids, polysaccharide, and other organic compounds react at various extents during fermentation and distillation processes; the polymerization most likely occurs in complex ways and extends to diverse levels. Accordingly, it is not easy to characterize those pigments responsible for color in DE.

vii. Understanding the structure and chemical characteristics of Maillard reaction products such as melanoidins are essential prior to the development of a suitable decolorization technique for DE.

DOI: 10.1201/9781003029885-14

viii. During degradation or microbial action, several compounds are also transformed into more noxious compounds in the environment. These compounds may also bind to other pollutants and convert them into more vulnerable chemical forms.

ix. The degradation and decolorization of melanoidins by microorganisms has been reported more effective due to the prevalence of extracellular ligninolytic enzymatic system known as lignin peroxidase (LiP), manganese-dependent peroxidase (MnP), and laccase (Lac), which are capable of breaking a large number of C=C, C=O, and C≡N bonds present in melanoidins. However, these enzyme-producing microorganisms do not completely metabolize melanoidins as a sole carbon and nitrogen sources.

x. During chlorine disinfection of DE, polyphenols and melanoidins may also lead to the development of aromatic halogenated disinfection by-products (DBPs), which are highly toxic and inhibit growth.

xi. The complexity of DE makes it difficult for the development of efficient tertiary treatment technologies for removal of color and toxic androgenic and mutagenic compounds before its disposal into the environment.

xii. Technical and financial constraints are two significant obstacles that have hindered waste management improvements in developing countries. In most developing countries, there is a serious lack of technical expertise as well as engineering infrastructure preventing the transition of open dumps to landfills.

xiii. Several initiatives, supported by the Department of Biotechnology, Government of India, are underway to develop bioremediation solutions for DE.

xiv. Most of the countries have stringent regulation and rigorous discharge norms for the disposal of agro-based industrial wastewaters. However, implementation of such norms is quite challenging in a few cases, which insists a steady infusion of cost-effective and innovative technologies with steady research efforts.

Mitigation of these challenges requires financial support to distilleries at large scale from the federal government for the upgradation of effluent treatment technologies. Thus, distillery wastewater has been under extensive research by the science community as it presents a serious danger to the aquatic and soil environment, especially with respect to human and animal health. Based on the knowledge generated up to now, several future avenues of research approaches have been suggested:

i. Researchers have prepared and tested different constituents simulating actual wastewater discharged from distilleries. However, the attention given to artificial DE remains limited.

ii. Many reports deal only with the treatment of DE with dilution of wastewater for obtaining better treatment efficiency or for successful experiments.

iii. Synthetic wastewater is necessary for application in pilot-scale studies and when real stable effluents are not available.

iv. Most of the decolorization and degradation studies of melanoidins are reported at 475 nm only based on purified melanoidin absorption maxima with specific molecular weight, while the DE contains a mixture of Maillard reaction products (MRPs), i.e., initial, intermediate, and advanced stages with variable molecular weight. Therefore, prior to attempting the decolorization and degradation of distillery effluent, the degradation of model melanoidins with a mixture of complex MRPs should be evaluated for its degradability.

v. In the context of DE degradation and decolorization, the majority of the research has been conducted in laboratories under relatively controlled conditions for short periods. To implement decolorization using microorganisms on a larger scale in the environment, more extensive research under field conditions for long durations is required for a better understanding of the potential role of bacteria or fungi in degradation and detoxification of DE pollutants, including melanoidins.

vi. DE shows variable absorption spectrum at a different wavelength due to the presence of a mixture of different MRPs. But absorption and elimination patterns of different peaks in melanoidins are not reported so far. Therefore, it is vital to reveal the decline pattern of the specific peak of melanoidins in degradation and generation of its metabolites formed during the decolorization process.

vii. The effect of sulfides and metals along with phenolic compounds required systematic studies to establish the mechanism of biological decolorization prior to its scope to develop an industrial-scale decolorization technique for safe disposal.

viii. Research is needed in order to understand the detailed mechanism of ligninolytic enzymes in degradation and/or biotransformation of melanoidins and other co-contaminants of DE.

ix. There is a need for more pilot and field studies to demonstrate the effectiveness of bacteria-based biological treatment technology and increase its acceptance.

x. Most results of microbial-based treatment were obtained from in vitro studies, but the fungal treatment processes may be influenced by various environmental factors, and more work should be carried out to document the role of naturally adapted indigenous fungi.

xi. There is no detailed data nor any adequate monitoring device for persistent organic pollutants discharged in effluent from distilleries. Hence, for screening, any program for remediation and monitoring of these pollutants at a regional/national level should be initiated for environmental protection and sustainable development.

xii. Not much data is yet available on the field performance of distillery effluent treatment using fungi/bacteria culture or consortium. Established field trials are, therefore, urgently needed to make it a commercially viable and acceptable technology.

xiii. Most studies on degradation and decolorization of DE are commonly based on experimental manipulations and are rarely based on variable and field-realistic conditions. Future research should move beyond these limitations and identify how microorganisms and their enzyme partnerships are working in tandem under different scenarios.

xiv. The applications of bacteria and wetland plants in two-step/sequential treatment have been reported to be a very promising technology for detoxification, but this has to be optimized yet with detailed physiology of wetland plants, microbiology of plant rhizospheres, and melanoidins detoxification mechanism along with co-pollutants.

xv. Microbial communities grown in such an adverse environment are very much important to understand the microbiology of distillery waste. There is a need for more investigation and research on the changes induced by microorganisms present in rhizosphere during phytoremediation of distillery waste pollutants.

xvi. More research is required to be performed on quantifying the effect of the phytoremediation process on the phytoavailability of melanoidins and toxic heavy metals.

xvii. There is also considerable scope for operational improvements in biomethanation of spent wash such as treating the spent wash without any dilution, ensuring shorter start-up periods, and degrading refractile components to improve the anaerobic treatment.

xviii. The examination of different pilot-scale treatment systems remains problematic due to the absence of a permanent source of real wastewater. Therefore, more attention needs to be paid to preparing synthetic wastewater, which simulates discharged effluents from distilleries.

xix. Weed plant appears to be a good choice for restoration of distillery waste contaminated site since these hardy, tolerant species can quickly grow in most harsh conditions, including heavy metals stress, over large areas and give a good amount of biomass as a secondary product.

xx. Weed plant can restrict the contaminant from being introduced into the food web and consequently reduce the risk to human health.

xxi. To reduce the potential ecological risk on local ecosystems posed by non-native weed species, more effective native weeds that are compatible with local habitats are preferred and need to be tested for use in phytoremediation.

xxii. The acquisition of heavy metals by plants depends on the ability of a plant and its rhizospheric microbial communities to solubilize and mobilize heavy metals in the rhizosphere. Research should thus be focused on increasing the population of PGPR in polluted soil and/or rhizo-sphere environments by inoculation of soils or plants.

xxiii. DE contains a large number of antioxidant compounds that could be recovered and serve as additional value-added resources.

xxiv. The integrated technology has the prospective to be developed into an environmentally and economically acceptable wastewater treatment technology for developing countries for sustainable development.

15

Concluding Remarks

Distilleries are one of the largest industrial sector generating vast quantities of effluent, which is potentially a great cause of aquatic and soil pollution. Thus, safe disposal of distillery waste into the environment and development of cost-effective and eco-friendly treatment techniques are urgently required for the efficient treatment of distillery effluent. The following conclusion can be made from this book:

i. Distillery effluent, a highly colored residue from the alcohol industry, is a major threat to the environment for its safe disposal due to their high biological oxygen demand (BOD), chemical oxygen demand (COD), total dissolved solids (TDS), unpleasant odors, and high content of various toxic and refractory inorganic and organic compounds, including heavy metals.

ii. The intense color of distillery effluent is mainly due to the existence of a dark brown polymeric pigment compound known as melanoidin, which exhibits antioxidant and antimicrobial properties that inhibit microbial growth used in conventional biological treatment processes of distillery effluent.

iii. Conventional biological treatment approaches has been accomplished the degradation and decolorization of melanoidins up to only 67%.

iv. The discharge of organic and inorganic pollutants into the aquatic and terrestrial ecosystem from distilleries creates a risk to organisms, causing significant environmental disruption.

v. In many regions of India, distillery industries dilute the biomethanated distillery effluent by mixing with raw water prior to discharge in the environment in order to meet the set effluent disposal standard. This dilution of effluent even though accepted in some regions of India is of great environmental concern as it does not reduce the complete pollution load of the spent wash.

vi. The two-stage aeration process presently in use leads to significant COD reduction, but it does not result in color removal.

vii. The large-scale applications of fungi-based treatment of distillery effluent have own constraint due to slow growth cycle, huge spore formation, low pH range (3–5), and adverse submerged aquatic environment for the growth of fungus.

viii. Phytoremediation is an eco-friendly, effective, solar-driven, and sustainable technology with good community acceptance compared to conventional biological and physicochemical treatment processes.

ix. Several potential weed plants that grow on distillery waste-contaminated sites under natural conditions have indicated the phytoremediation potential of the use of such plants at contaminated sites as natural hyperaccumulators of heavy metals from complex organometallic wastes.

x. Phytoextraction is a polishing green approach wherein the action of plants and the microorganisms specifically associated with them is optimized to sequester, degrade, transform, assimilate, metabolize, or detoxify hazardous pollutants from the distillery sludge as well as contaminated site.

xi. The commercialization of phytoremediation is lagging behind, as less information is available about treatment success at pilot/field scale. The need of the time is to get back the industry's confidence by developing cost-effective and environment-friendly solutions for the on-site treatment of distillery waste.

DOI: 10.1201/9781003029885-15

xii. The biphasic/phase separation/sequential treatment system using microorganisms and constructed wetland plant system is a practical, environment-friendly, and economical substitute technology to replace traditional secondary natural ecosystem for treating distillery wastewater.

xiii. Also given the low priority allocated to waste management, very limited funds are provided to the distillery waste management sector by the governments. The funds are often not sufficient to achieve the level of protection required for public health and the environment.

xiv. The complexity of distillery effluent makes it difficult for the development of efficient tertiary treatment technologies for removal of color and toxic androgenic and mutagenic compounds prior to its disposal into the environment.

xv. Up to the present, however, no suitable technique or method for the complete treatment of large amounts of distillery effluent has been developed.

References

Acharya, B.K., Pathak, H., Mohana, S., Shouche, Y., Singh, V., & Madamwar, D. (2011). Kinetic modelling and microbial community assessment of anaerobic biphasic fixed film bioreactor treating distillery spent wash. *Water Research* 45(14), 4248–4259.

Adikane, H.V., Dange, M.N., & Selvakumari, K. (2006). Optimization of anaerobically digested distillery molasses spent wash decolorization using soil as inoculum in the absence of additional carbon and nitrogen source. *Bioresource Technology* 97, 2131–213.

Ahansazan, B., Afrashteh, H., Ahansazan, N., & Ahansazan, Z. (2014). Activated sludge process overview. *International Journal of Environmental Science and Development* 5(1), 81–85.

Ahmed, P.M., Pajot, H.F., de Figueroa, L.I.C., & Gusils, C.H. (2018). Sustainable bioremediation of sugarcane vinasse using autochthonous macrofungi. *Journal of Environmental Chemical Engineering* 6(4), 5177–5185.

Ahmed, S., Unar, I.N., Khan, H.A., Maitlo, G., Mahar, R.B., Jatoi, A.S., Memon, A.Q., & Shah, A.K. (2020). Experimental study and dynamic simulation of melanoidin adsorption from distillery effluent. *Environmental Science and Pollution Research* 27, 9619–9636.

Alves, P.R.L., Natal-da-Luz, T., Sousa, J.P., & Cardoso, E.J.B.N. (2015). Ecotoxicological characterization of sugarcane vinasses when applied to tropical soils. *Science of the Total Environment* 526, 222–232.

Amenorfenyo, D.K., Huang, X., Zhang, Y., Zeng, Q., Zhang, N., Ren, J., & Huang, Q. (2019). Microalgae brewery wastewater treatment: potentials, benefits and the challenges. *International Journal of Environmental Research and Public Health* 16, 1910.

Anbarasan, T., Jayanthi, S., & Ragina, Y., 2018. Investigation on synthesis of biodiesel from distillery spent wash using oleaginous yeast *Metschnikowia Pulcherrima*. *Materialstoday Proceedings* 5(11), 23293–23301.

Angayarkanni, J., Palaniswamy, M., & Swaminathan, K. (2003). Biotreatment of distillery effluent using *Aspergillus niveus. Bulletin of Environmental Contamination and Toxicology* 70(2), 268–277.

Apollo, S., & Aoyi, O. (2016). Combined anaerobic digestion and photocatalytic treatment of distillery effluent in fluidized bed reactors focusing on energy conservation. *Environmental Technology (United Kingdom)* 37(17), 2243–2251.

Aragão, M.S., Menezes, D.B., Ramos, L.C., Oliveira, H.S., Bharagava, N., Ferreira, L.F.R., Teixeira, J.A., Ruzene, D.S., & Silva, D.P. (2019). Mycoremediation of vinasse by surface response methodology and preliminary studies in air-lift bioreactors. *Chemosphere* 244, 125432.

Arimi, M.M. (2017). Modified natural zeolite as heterogeneous Fenton catalyst in treatment of recalcitrants in industrial effluent. *Progress in Natural Science: Materials International* 27(2), 275–282.

Arimi, M.M., Zhang, Y., & Geißen, S.U. (2015). Color removal of melanoidin-rich industrial effluent by natural manganese oxides. *Separation and Purification Technology* 150, 286–291.

Arimi, M.M., Zhang, Y., Götz, G., Kiriamiti, K., & Geißen, S. (2014). Antimicrobial colorants in molasses distillery wastewater and their removal technologies. *International Biodeterioration & Biodegradation* 87, 34–43.

Arora, M., Sharma, D.K., & Behera, B.K. (1992). Upgrading of distillery effluent by *Nitrosococcus oceanus* for its use as a low-cost fertilizer. *Resources, Conservation and Recycling* 6(4), 347–353.

Asaithambi, P., Garlanka, L., Anantharaman, N., & Matheswaran, M. (2012). Influence of experimental parameters in the treatment of distillery effluent by electrochemical oxidation. *Separation Science and Technology* 47(3), 470–481.

Asaithambi, P., Govindarajan, R., Yesuf, M.B., Selvakumar, P., & Alemayehu, E. (2021). Investigation of direct and alternating current–electrocoagulation process for the treatment of distillery industrial effluent: Studies on operating parameters. *Journal of Environmental Chemical Engineering* 9(2), 104811.

Asano, R., Kobayashi, S., Sonobe, K., Shime-Hattori, A., Okazaki, K., & Ohtomo, R. (2014). Plant-available inorganic nutrient levels are increased in rice-derived distillery effluents inoculated with microbes. *Journal of Applied Microbiology* 117(5), 1412–1421.

Ayyasamy, P.M., Yasodha, R., Rajakumar, S., Lakshmanaperumalsamy, P., Rahman, P.K.S.M., & Lee, S. (2008). Impact of sugar factory effluent on the growth and biochemical characteristics of terrestrial and aquatic plants. *Bulletin of Environmental Contamination and Toxicology* 81(5), 449–454.

Bachmann, A. (1983). Comparison of fixed film reactors with a modified sludge blanket reactor. *Pollution Technology Review* 10, 384–402.

Barakat, M.A. (2011). New trends in removing heavy metals from industrial wastewater. *Arabian Journal of Chemistry* 4(4), 361–377.

Bento, I., Silva, C.S., Chen, Z., Martins, L.O., & Lindley, P.F., & Soares, C.M. (2010). Mechanisms underlying dioxygen reduction in laccases. structural and modelling studies focusing on proton transfer. *BMC Structural Biology* 10, 28. https://doi.org/10.1186/1472-6807-10-28

Bharagava, R.N., & Chandra, R. (2010). Biodegradation of the major color containing compounds in distillery wastewater by an aerobic bacterial culture and characterization of their metabolites. *Biodegradation* 21(5), 703–711.

Bharagava, R.N., Chandra, R., & Rai, V. (2008). Phytoextraction of trace elements and physiological changes in Indian mustard plants (*Brassica nigra* L.) grown in post methanated distillery effluent (PMDE) irrigated soil. *Bioresource Technology* 99, 8316–8324.

Bharagava, R.N., Chandra, R., & Rai, V. (2009). Isolation and characterization of aerobic bacteria capable of the degradation of synthetic and natural melanoidins from distillery effluent. *World Journal of Microbiology and Biotechnology* 25(5), 737–744.

Bhoite, G.M., & Vaidya, P.D. (2018a). Improved biogas generation from biomethanated distillery wastewater by pretreatment with catalytic wet air oxidation. *Industrial and Engineering Chemistry Research* 57(7), 2698–2704.

Bhoite, G.M., & Vaidya, P.D. (2018b). Iron-catalyzed wet air oxidation of biomethanated distillery wastewater for enhanced biogas recovery. *Journal of Environmental Management* 226, 241–248.

Billore, S.K., Singh, N., Ram, H.K., Sharma, J.K., Singh, V.P., Nelson, R.M., & Dass, P. (2001). Treatment of a molasses-based distillery effluent in a constructed wetland in central India. *Water Science and Technology* 44(11–12), 441–448.

Blonskaja, V., Menert, A., Vilu, R. (2003). Use of two-stage anaerobic treatment for distillery waste. *Advances in Environmental Research* 7, 671–678.

Blonskaja, V., & Zub, S. (2009). Possible ways for post-treatment of biologically treated wastewater from yeast factory. *Journal of Environmental Engineering and Landscape Management* 17(4), 189–197.

Boczkaj, G., & Fernandes, A. (2017). Wastewater treatment by means of advanced oxidation processes at basic pH conditions: A review. *Chemical Engineering Journal* 320, 608–633.

Borja, R., & Banks, C. (1994). Anaerobic digestion of palm oil mill effluent using an up-flow anaerobic sludge blanket (UASB) reactor. *Biomass and Bioenergy* 6(5), 381–389.

Caderby, E., Baumberger, S., Hoareau, W., Fargues, C., Decloux, M., & Maillard, M.N. (2013). Sugar cane stillage: A potential source of natural antioxidants. *Journal of Agricultural and Food Chemistry* 61(47), 11494–11501.

Cammerer, B., Jaluschkov, V., & Kroh, L.W. (2002). Carbohydrates structures as part of the melanoidins skeleton. *International Congress Series* 1245, 269.

Cao, Y., Mu, H., Liu, W., Zhang, R., Guo, J., Xian, M., & Liu, H. (2019). Electricigens in the anode of microbial fuel cells: Pure cultures versus mixed communities. *Microbial Cell Factories* 18, 39.

Cassman, N.A., Lourenço, K.S., do Carmo, J.B., Cantarella, H., & Kuramae, E.E. (2018). Genome-resolved metagenomics of sugarcane vinasse bacteria. *Biotechnology for Biofuels* 11, 48.

Chairattanamanokorn, P., Imai, T., Kondo, R., Sekine, M., Higuchi, T., & Ukita, M. (2005). Decolorization of alcohol distillery wastewater by thermotolerant white-rot fungi. *Prikladnaia Biokhimiiai Mikrobiologiia* 41(6), 662–667.

Chandra, R., Bharagava, R.N., Kapley, A., & Purohit, H.J. (2012). Characterization of *Phragmites cummunis* rhizosphere bacterial communities and metabolic products during the two stage sequential treatment of post methanated distillery effluent by bacteria and wetland plants. *Bioresource Technology* 103(1), 78–86.

Chandra, R., Bharagava, R.N., & Rai, V. (2008). Melanoidins as major colourant in sugarcane molasses based distillery effluent and its degradation. *Bioresource Technology* 99(11), 4648–4660.

Chandra, R., Bharagava, R.N., Rai, V., & Singh, S.K. (2009). Characterization of sucrose-glutamic acid Maillard products (SGMPs) degrading bacteria and their metabolites. *Bioresource Technology* 100(24), 6665–6668.

Chandra, R., Dubey, N.K., & Kumar, V. (2018d) (Eds), *Phytoremediation of environmental pollutants.* CRC Press: Boca Raton

Chandra, R., & Kumar, V. (2015a) Biotransformation and biodegradation of organophosphates and organo-halides. In: Chandra, R. (Ed), *Environmental waste management.* CRC Press: Boca Raton.

Chandra, R., & Kumar, V. (2015b) Mechanism of wetland plant rhizosphere bacteria for bioremediation of pollutants in an aquatic ecosystem. In: Chandra, R. (Ed), *Advances in biodegradation and bioremediation of industrial waste.* CRC Press: Boca Raton.

Chandra, R., & Kumar, V. (2017a). Detection of *Bacillus* and *Stenotrophomonas* species growing in an organic acid and endocrine-disrupting chemical-rich environment of distillery spent wash and its phytotoxicity. *Environmental Monitoring and Assessment* 189(1), 26.

Chandra, R., & Kumar, V. (2017b). Detection of androgenic-mutagenic compounds and potential autochthonous bacterial communities during in situ bioremediation of post-methanated distillery sludge. *Frontiers in Microbiology* 8, 87.

Chandra, R., & Kumar, V. (2017c). Phytoextraction of heavy metals by potential native plants and their microscopic observation of root growing on stabilised distillery sludge as a prospective tool for in situ phytoremediation of industrial waste. *Environmental Science and Pollution Research* 24, 2605–2619.

Chandra, R., & Kumar, V. (2018). Phytoremediation: A green sustainable technology for industrial waste management. In: Chandra, R., Dubey, N.K., Kumar, V. (Eds), *Phytoremediation of environmental pollutants.* CRC Press: Boca Raton.

Chandra, R., Kumar, V., & Tripathi, S. (2018a). Evaluation of molasses-melanoidin decolourisation by potential bacterial consortium discharged in distillery effluent. *3 Biotech* 8, 187.

Chandra, R., Kumar, V., Tripathi, S., & Sharma, P. (2018b). Heavy metal phytoextraction potential of native weeds and grasses from endocrine-disrupting chemicals rich complex distillery sludge and their histological observations during in-situ phytoremediation. *Ecological Engineering* 111, 143–156.

Chandra, R., Kumar, V., & Singh, K. (2018c). Hyperaccumulator versus nonhyperaccumulator plants for environmental waste management. In: Chandra, R., Dubey, N.K., Kumar, V. (Eds), *Phytoremediation of environmental pollutants.* CRC Press: Boca Raton.

Chandra, R., Kumar, V., & Yadav, S. (2017). Extremophilic ligninolytic enzymes. In: Sani R., Krishnaraj R. (Eds), *Extremophilic enzymatic processing of lignocellulosic feedstocks to bioenergy.* Springer: Cham. https://doi.org/10.1007/978-3-319-54684-1_8

Chandra, R., Saxena, G., & Kumar, V. (2015) Phytoremediation of environmental pollutants: An eco-sustainable green technology to environmental management. In: Chandra, R. (Ed), *Advances in biodegradation and bioremediation of industrial waste.* CRC Press: Boca Raton.

Chandra, R., & Yadav, S. (2010). Potential of *Typha angustifolia* for phytoremediation of heavy metals from aqueous solution of phenol and melanoidin. *Ecological Engineering* 36(10), 1277–1284.

Chaturvedi, S., Chandra, R., & Rai, V. (2006). Isolation and characterization of *Phragmites australis* (L.) rhizosphere bacteria from contaminated site for bioremediation of colored distillery effluent. *Ecological Engineering* 27(3), 202–207.

Chaturvedi, V., & Verma, P. (2016). Microbial fuel cell: A green approach for the utilization of waste for the generation of bioelectricity. *Bioresource and Bioprocessing* 3, 38.

Chauhan, J., & Rai, J.P.N. (2010). Monitoring of impact of ferti-irrigation by postmethanated distillery effluent on groundwater quality. *Clean – Soil, Air, Water* 38(7), 630–638.

Chavan, M.N., Dandi, N.D., Kulkarni, M.V., & Chaudhari, A.B. (2013). Biotreatment of melanoidin-containing distillery spent wash effluent by free and immobilized *Aspergillus oryzae* MTCC 7691. *Water, Air, and Soil Pollution* 224(11), 1755.

Chavan, M.N., Kulkarni, V., Zope, V.P., & Mahulikar, P.P. (2006). Microbial degradation of melanoidins in distillery spent wash by an indigenous isolate. *Indian Journal of Biotechnology* 5, 416–421.

Chin Fa, H., Ying Shu, J., Shyang Chwen, S., Pao Chuan, H., & Jia Hsin, G. (2011). Purification and characterization of a novel glucose oxidase-like melanoidin decolorizing enzyme from *Geotrichum* sp. No. 56. *African Journal of Microbiology Research* 5(20), 3256–3266.

Chinnasamy, S., Bhatnagar, A., Hunt, R.W., & Das, K.C. (2010). Microalgae cultivation in a wastewater dominated by carpet mill effluents for biofuel applications. *Bioresource Technology* 101(9), 3097–3105.

Chopra, P., Singh, D., Verma, V., Puniya, A.K., 2004. Bioremediation of melanoidins containing digested spent wash from cane-molasses distillery with white rot fungus, Coriolus versicolour. *Indian Journal of Microbiology* 44, 197–200.

Chowdhary, P., Raj, A., Bharagava, R.N. (2018a). Environmental pollution and health hazards from distillery wastewater and treatment approaches to combat the environmental. *Chemosphere* 194, 229–246.

Chowdhary, P., Sammi, S.R., Pandey, R., Kaithwas, G., Raj, A., Singh, J., & Bharagava, R.N. (2020). Bacterial degradation of distillery wastewater pollutants and their metabolites characterization and its toxicity evaluation by using *Caenorhabditis elegans* as terrestrial test models. *Chemosphere* 261, 127689.

Chowdhary, P., Yadav, A., Singh, R., Chandra, R., Singh, D. P., Raj, A., & Bharagava, R. N. (2018b) Stress response of *Triticum aestivum* L. and *Brassica juncea* L. against heavy metals growing at distillery and tannery wastewater contaminated site. *Chemosphere* 206, 122–131.

Christofoletti, C.A., Escher, J.P., Correia, J.E., Marinho, J.F.U., &Fontanetti, C.S. (2013). Sugarcane vinasse: environmental implications of its use. *Waste Management* 33, 2752–2761.

Chuppa-Tostain, G., Hoarau, J., Watson, M., Adelard, L., Shum Cheong Sing, A., Caro, Y., Grondin, I., Bourven, I., Francois, J.-M., Girbal-Neuhauser, E., & Petit, T. (2018). Production of *Aspergillus niger* biomass on sugarcane distillery wastewater: Physiological aspects and potential for biodiesel production. *Fungal Biology and Biotechnology* 5(1), 1–12.

Cortez, L., Magalhães, P., & Happi, J. (1992). Principais subprodutos da agroindústria canavieira e sua valorização. *Brazilian Energy Magazine* 2, 1–17.

CPCB (Central Pollution Control Board). (2003). Annual report; 2002–2003. http://www.cpcb.nic.in/ar2003/ar2-3ch10.htm. CPCB (Central Pollution Control Board). Chapter on corporate responsibility for environmental protection, distillery.

D'souza, D.T., Tiwari, R., Sah, A.H., & Raghukumar, C. (2006). Enhanced production of laccase by a marine fungus during treatment of colored effluents and synthetic dyes. *Enzyme and Microbial Technology* 38, 504–511.

Dahiya, J., Singh, D., & Nigam, P. (2001). Decolourisation of synthetic and spentwash melanoidins using the white-rot fungus *Phanerochaete chrysosporium* JAG-40. *Bioresource Technology* 78, 95–98.

Dashtban, M., Schraft, H., Syed, T.A., Qin, W., 2010. Fungal biodegradation and enzymatic modification of lignin. *International Journal of Biochemistry and Molecular Biology* 1(1), 36.

Davamani, V., Lourduraj, A.C., & Singaram, P. (2006). Effect of sugar and distillery wastes on nutrient status, yield and quality of turmeric. *Crop Research Hisar* 32(3), 563–567.

David, C., Arivazhagan, M., & Ibrahim, M. (2015a). Spent wash decolourization using nano-Al_2O_3/kaolin photocatalyst: Taguchi and ANN approach. *Journal of Saudi Chemical Society* 19(5), 537–548.

David, C., Arivazhagan, M., Balamurali, M.N., & Shanmugarajan, D. (2015b). Decolorization of distillery spent wash using biopolymer synthesized by *Pseudomonas aeruginosa* isolated from tannery effluent. *BioMed Research International* 2015.

David, C., Arivazhagan, M., Hariram, J., & Sruthi, P. (2016). Spent wash decolorization using granular and powdered activated carbon: Taguchi's orthogonal array design and ANN approach. *Journal of Applied Science and Engineering Methodologies* 2(2), 224–229.

David, C., Arivazhagann, M., & Tuvakara, F. (2015c). Decolorization of distillery spent wash effluent by electrooxidation (EC and EF) and Fenton processes: A comparative study. *Ecotoxicology and Environmental Safety* 121, 142–148.

David, C., Narlawar, R., & Arivazhagan, M. (2016). Performance evaluation of moringa oleifera seed extract (MOSE) in conjunction with chemical coagulants for treating distillery spent wash. *Indian Chemical Engineer* 58(3), 189–200.

de Barros, V.G., Duda, R.M., da Silva Vantini, J., Omori, W.P., Ferro, M.I.T., & de Oliveira, R.A. (2017). Improved methane production from sugarcane vinasse with filter cake in thermophilic UASB reactors, with predominance of *Methanothermobacter* and *Methanosarcina* archaea and *Thermotogae* bacteria. *Bioresource Technology* 244(Part 1), 371–381.

Deng, Y., & Zhao, R. (2015). Advanced oxidation processes (AOPs) in wastewater treatment. *Current Pollution Reports* 1(3), 167–176.

Derakhshan, M., & Fazeli, M. (2018). Improved biodegradability of hardly-decomposable wastewaters from petrochemical industry through photo-Fenton method and determination of optimum operational conditions by response surface methodology. *Journal of Biological Engineering* 12, 10.

Devarajan, L., Rajanan, G., Ramanathan, G., & Oblisami, G. (1994). Performance of field crops under distillery effluent irrigations. *Kisan World* 21, 48–50

Driessen, W., Tielbaard, M., & Vereijken, T. (1994). Experience on anaerobic treatment of distillery effluent with the UASB process. *Water Science and Technology* 30(12), 193–201.

Echavarría, A.P., Pagán, J., & Ibarz, A. (2012). Melanoidins formed by Maillard reaction in food and their biological activity. *Food Engineering Reviews* 4(4), 203–223.

España-Gamboa, E., Vicent, T., Font, X., Dominguez-Maldonado, J., Canto-Canché, B., & Alzate-Gaviria, L. (2017). Pretreatment of vinasse from the sugar refinery industry under non-sterile conditions by *Trametes versicolor* in a fluidized bed bioreactor and its effect when coupled to an UASB reactor. *Journal of Biological Engineering* 11(1), 1–11.

España-Gamboa, E., Vicent, T., Font, X., Mijangos-Cortés, J., Canto-Canché, B., & Alzate, L. (2015). Phenol and color removal in hydrous ethanol vinasses in an air-pulsed bioreactor using *Trametes versicolor*. *Journal of Biochemical Technology* 6, 982–986.

Fahy, V., FitzGibbon, F.J., McMullan, G., Singh, D., & Marchant, R. (1997). Decolourisation of molasses spent wash by *Phanerochaete chrysosporium*. *Biotechnology Letters* 19(1), 97–99.

Fan, L., Fan, L., Nguyen, T., & Roddick, F.A. (2011). Characterisation of the impact of coagulation and anaerobic bio-treatment on the removal of chromophores from molasses wastewater bio-treatment on the removal of chromophores from molasses wastewater. *Water Research* 45(13), 3933–3940.

Fernandez-Fueyo, E., Ruiz-Duenas, F.J., Martinez, M.J., Romero, A., Hammel, K.E., Medrano, F.J., & Martinez, A.T. (2014). Ligninolytic peroxidase genes in the oyster mushroom genome: Heterologous expression, molecular structure, catalytic and stability properties, and lignin-degrading ability. *Biotechnology for Biofuels* 7, 2.

Ferreira, L.F., Aguiar, M., Pompeu, G., Messias, T.G., & Monteiro, R.R. (2010). Selection of vinasse degrading microorganisms. *World Journal of Microbiology and Biotechnology* 26(9), 1613–1621.

Ferreira, L.F.R., Aguiar, M. M., Messias, T.G., Pompeu, G.B., Lopez, A.M.Q., Silva, D.P., & Monteiro, R.T. (2011). Evaluation of sugar-cane vinasse treated with *Pleurotus sajor-caju* utilizing aquatic organisms as toxicological indicators. *Ecotoxicology and Environmental Safety* 74(1), 132–137.

Fito, J., Tefera, N., Kloos, H., & Van Hulle, S.W.H. (2018). Physicochemical properties of the sugar industry and ethanol distillery wastewater and their impact on the environment. *Sugar Technology* 21, 265–277.

Fito, J., Tefera, N., Van Hulle, & S.W. H. (2019). Sugarcane biorefineries wastewater: bioremediation technologies for environmental sustainability. *Chemical and Biological Technologies in Agriculture* 6, 6. https://doi.org/10.1186/s40538-019-0144-5

FitzGibbon, F., Singh, D., McMullan, G., & Marchant, R. (1998). The effect of phenolic acids and molasses spent wash concentration on distillery wastewater remediation by fungi. *Process Biochemistry* 33(8), 799–803, ISSN 1359-5113.

Francisca Kalavathi, D., Uma, L., & Subramanian, G. (2001). Degradation and metabolization of the pigment-melanoidin in distillery effluent by the marine cyanobacterium *Oscillatoria boryana* BDU 92181. *Enzyme and Microbial Technology* 29(4–5), 246–251.

Gayathri, G., & Srinikethan, G. (2019). Bacterial cellulose production by K. Saccharivorans BC1 strain using crude distillery effluent as cheap and cost effective nutrient medium. *International Journal of Biological Macromolecules* 138, 950–957.

Georgiou, R.P., Tsiakiri, E.P., Lazaridis, N.K., & Pantazaki, A.A. (2016). Decolorization of melanoidins from simulated and industrial molasses effluents by immobilized laccase. *Journal of Environmental Chemical Engineering* 4(1), 1322–1331.

Georgiou, R., Tsiakiri, E.P., & Pantazaki, A.A. (2014). Immobilization of laccase on alumina or controlled pore glass-uncoated nanoparticles and decolorization of melanoidin from bakery effluents. International Congress on Water, Waste and Energy Management, Porto Portugal, 16–18 July, Abstracts Book EWWM2014, p. 90.

Ghosh, M., Ganguli, A., & Tripathi, A.K. (2002). Treatment of anaerobically digested distillery spent wash in a two-stage bioreactor using *Pseudomonas putida* and *Aeromonas* sp. *Process Biochemistry* 37, 857–862.

Ghosh, M., Verma, S.C., Mengoni, A., & Tripathi, A.K. (2004). Enrichment and identification of bacteria capable of reducing chemical oxygen demand of anaerobically treated molasses spent wash. *Journal of Applied Microbiology* 96(6), 1278–1286.

Ghosh Ray, S., & Ghangrekar, M.M. (2015). Enhancing organic matter removal, biopolymer recovery and electricity generation from distillery wastewater by combining fungal fermentation and microbial fuel cell. *Bioresource Technology* 176, 8–14.

Godshall, M. (1996). Recent progress in sugar colourants: GC-MS. *Proceedings of the South African Sugar Technologist's Association* 70, 153–161.

González, T., Terrón, M.C., Yagüe, S., Zapico, E., Galletti, G.C., & González, A.E. (2000). Pyrolysis/gas chromatography/mass spectrometry monitoring of fungal- biotreated distillery wastewater using *Trametes* sp. I-62 (CECT 20197). *Rapid Communications in Mass Spectrometry* 14(15), 1417–1424.

González, T., Terrón, M.C., Yagüe, S., Junca, H., Carbajo, J.M., Zapico, E.J., Silva, R., Arana-Cuenca, A., Téllez, A., & González, A.E. (2008). Melanoidin-containing wastewaters induce selective laccase gene expression in the white-rot fungus *Trametes* sp. I-62. *Research in Microbiology* 159(2), 103–109.

Goodwin, J.A.S., Finlayson, J.M., & Low, E.W. (2001). A further study of the anaerobic biotreatment of malt whisky distillery pot ale using an UASB system. *Bioresource Technology* 78(2), 155–160

Guan, Z.B., Zhang, N., Song,C.-M., Zhou, W., Zhou, L.-X., Zhao, H., Xu, C.-W., Cai, Y.-J., & Liao, X.-R. (2013). Molecular cloning, characterization, and dye-decolorizing ability of a temperature- and ph-stable laccase from *Bacillus subtilis* X1. *Applied Biochemistry and Biotechnology* 172, 1147–1157. https://doi.org/10.1007/s12010-013-0614-3

Guimarães, C., San Miguel Bento, L., & Mota, M. (1999). Biodegradation of colorants in refinery effluents: Potential use of the fungus *Phanerochaete chrysosporium*. *International Sugar Journal* 101(1205), 246–251.

Ha, P.T., Lee, T.K., Rittmann, B.E., Park, J., & Chang, I.S. (2012). Treatment of alcohol distillery wastewater using a bacteroidetes-dominant thermophilic microbial fuel cell. *Environmental Science and Technology* 46(5), 3022–3030.

Harada, H. (1996). Anaerobic treatment of a recalcitrant distillery wastewater by a thermophilic UASB. *Bioresource Technology* 55(3), 215–221.

Harada, H., Uemura, S., Chen, A.C., & Jayadevan, J. (1996). Anaerobic treatment of a recalcitrant distillery wastewater by a thermophilic UASB reactor. *Bioresource Technology* 55, 215–221.

Hatano, K.-I., Kanazawa, K., Tomura, H., Yamatsu, T., Tsunoda, K.-I., & Kubota, K. (2016). Molasses melanoidin promotes copper uptake for radish sprouts: The potential for an accelerator of phytoextraction. *Environmental Science and Pollution Research* 23(17), 17656–17663.

Hatano, K.I., & Yamatsu, T. (2018). Molasses melanoidin-like products enhance phytoextraction of lead through three *Brassica* species. *International Journal of Phytoremediation* 20(6), 552–559.

Hati, K.M., Biswas, A.K., Bandyopadhyay, K., & Misra, A.K. (2004). Effect of post-methanation effluent on soil physical properties under a soybean-wheat system in a Vertisol. *Journal of Plant Nutrition and Soil Science* 167(5), 584–590.

Hayase, F., Kim, S.B., & Kato, H. (1984). Decolorization and degradation products of the melanoidins by hydrogen peroxide. *Agricultural and Biological Chemistry* 48(11), 2711–2717.

Hayase, F. (2000). Recent development of 3-deoxyosone related Maillard reaction products. *Food Science and Technology Research* 6(2), 79–86.

Hemalatha, B. (2012). Recycling of industrial sludge along with municipal solid waste vermicomposting method. *International Journal of Advanced Engineering Technology* 3(2), 71–74.

Hoarau, J., Grondin, I., Caro, Y., & Petit, T. (2018). Sugarcane distillery spent wash, a new resource for third-generation biodiesel production. *Water (Switzerland)* 11, 1623.

Hoarau, J., Petit, T., Grondin, I., Marty, A., & Caro, Y. (2020). Phosphate as a limiting factor for the improvement of single cell oil production from *Yarrowia lipolytica* MUCL 30108 grown on pre-treated distillery spent wash. *Journal of Water Process Engineering* 37, 101392.

Hodge, J.E. (1953). Dehydrated foods. Chemistry of browning reactions in model systems. *Journal of Agricultural and Food Chemistry* 1, 928–943.

Hofmann, T. (1998). Studies on the relationship between molecular weight and the color potency of fractions obtained by thermal treatment of glucose/amino acid and glucose/protein solutions by using ultracentrifugation and color dilution techniques. *Journal of Agricultural and Food Chemistry* 46(10), 3891–3895.

Ince, O., Kolukirik, M., Oz, N.A., & Ince, B.K. (2005). Comparative evaluation of full-scale UASB reactors treating alcohol distillery wastewaters in terms of performance and methanogenic activity. *Journal of Chemical Technology and Biotechnology* 80, 138–144.

Insoongnoen, B., Yimrattanabovorn, J., & Wichitsathian, B. (2020). The potential of activated carbon production from *Leucaena leucocephala* charcoal and application in melanoidin removal. *Naresuan University Journal: Science and Technology* 28(1), 10–22.

Jack, F., Bostock, J., Tito, D., Harrison, B., & Brosnan, J. (2014). Electrocoagulation for the removal of copper from distillery waste streams. *Journal of the Institute of Brewing* 120(1), 60–64.

Jahan, F., Kumar, V., & Saxena, R.K. (2018). Distillery effluent as a potential medium for bacterial cellulose production: A biopolymer of great commercial importance. *Bioresource Technology* 250, 922–926.

Jain, N., Bhatia, A., Kaushik, R., Kumar, S., Joshi, H.C., & Pathak, H. (2005). Impact of post-methanation distillery effluent irrigation on groundwater quality. *Environmental Monitoring and Assessment* 110(1–3), 243–255.

Jain, R., & Srivastava, S. (2012). Nutrient composition of spent wash and its impact on sugarcane growth and biochemical attributes. *Physiology and Molecular Biology of Plants* 18(1), 95–99.

Jiménez, A.M., & Borja, R. (1997). Influence of aerobic pretreatment with *Penicillium decumbens* on the anaerobic digestion of beet molasses alcoholic fermentation wastewater in suspended and immobilized cell bioreactors. *Journal of Chemical Technology and Biotechnology* 69, 193–202.

Jiménez, A.M., Borja, R., & Martín, A. (2003). Aerobic-anaerobic biodegradation of beet molasses alcoholic fermentation wastewater. *Process Biochemistry* 38(9), 1275–1284.

Jiranuntipon, S., Chareonpornwattana, S., Damronglerd, S., Albasi, C., & Delia, M.L. (2008). Decolorization of synthetic melanoidins-containing wastewater by a bacterial consortium. *Journal of Industrial Microbiology and Biotechnology* 35(11), 1313–1321.

Jiranuntipon, S., Delia, M.L., Albasi, C., Damronglerd, S., & Chareonpornwattana, S. (2009). Decolourization of molasses based distillery wastewater using a bacterial consortium. *Science Asia* 35(4), 332–339.

Junior, J.A., Vieira, Y.A., Cruz, I.A., da Silva Vilar, D., Aguiar, M.M., Torres, N.H., Bharagava, R.N., Lima, Á.S., de Souza, R.L., & Romanholo Ferreira, L.F. (2020). Sequential degradation of raw vinasse by a laccase enzyme producing fungus *Pleurotus sajor-caju* and its ATPS purification. *Biotechnology Reports* 25, e00411.

Juwarkar, A., & Dutta, S. A. (1989). Impact of distillery effluent application to land on soil microflora. *Environmental Monitoring and Assessment* 15, 201–210.

Kahraman, S., & Yesilada, O. (2003). Decolorization and bioremediation of molasses wastewater by white rot fungi in a semi-solid-state condition. *Folia Microbiol* 48, 525–528.

Kallio, J.P., Gasparetti, C., Andberg, M., Boer, H., Koivula, A., Kruus, K., Rouvinen, J., & Hakulinen, N. (2010). Crystal structure of an ascomycete fungal laccase from *Thielavia arenaria* common structural features of asco-laccases. *BMC Structural Biology* 10, 29.

Kaparaju, P., Serrano, M., & Angelidaki, I. (2010). Optimization of biogas production from wheat straw stillage in UASB reactor. *Applied Energy* 87, 3779–3783.

Kaushik, A., Basu, S., Raturi, S., Batra, V.S., & Balakrishnan, M. (2018). Recovery of antioxidants from sugarcane molasses distillery wastewater and its effect on biomethanation. *Journal of Water Process Engineering* 25(August), 205–211.

Kaushik, A., Nisha, R., Jagjeeta, K., & Kaushik, C.P. (2005). Impact of long and short term irrigation of a sodic soil with distillery effluent in combination with bioamendments. *Bioresource Technology* 96(17), 1860–1866.

Kaushik, G., Gopal, M., & Thakur, I.S. (2010). Evaluation of performance and community dynamics of microorganisms during treatment of distillery spent wash in a three stage bioreactor. *Bioresource Technology* 101(12), 4296–4305.

Kaushik, G., & Thakur, I.S. (2009a). Isolation of fungi and optimization of process parameters for decolorization of distillery mill effluent. *World Journal of Microbiology and Biotechnology* 25(6), 955–964.

Kaushik, G., & Thakur, I.S. (2009b). Isolation and characterization of distillery spent wash color reducing bacteria and process optimization by Taguchi approach. *International Biodeterioration & Biodegradation* 63(4), 420–426.

Kaushik, G., & Thakur, I.S. (2010). Evaluation of performance and community dynamics of microorganisms during treatment of distillery spent wash in a three stage bioreactor. *Bioresource Technology* 101, 4296–4305.

Kaushik, G., & Thakur, I.S. (2013). Adsorption of colored pollutants from distillery spent wash by untreated and treated fungus: *Neurospora intermedia*. *Environmental Science and Pollution Research* 20, 1070–1078.

Kaushik, G., & Thakur, I.S. (2014). Production of laccase and optimization of its production by Bacillus sp. using distillery spent wash as inducer. *Bioremediation Journal* 18(1), 28–37.

Kehrein, P., & Garfí, M. (2020). Municipal wastewater treatment plants – Market. *Environmental Science Water Research & Technology*, 877–910.

Kelley, R.L., & Reddy, C.A. (1986). Identification of glucose oxidase activity as the primary source of hydrogen peroxide production in ligninolytic cultures of *Phanerochaete chrysosporium*. *Achieve in Microbiology* 144, 248–253.

Kidgell, J.T., De Nys, R., Hu, Y., Paul, N.A., & Roberts, D.A. (2014). Bioremediation of a complex industrial effluent by biosorbents derived from freshwater macroalgae. *PLoS ONE*, 9(6). 94706.

Kim, S.B., Hayase, F., & Kim, H. (1985). Decolorization and degradation products of melanoidins on ozonolysis. *Agricultural and Biological Chemistry* 49(3), 785–792.

Kim, S.J., & Shoda, M. (1999). Purification and characterization of a novel peroxidase from Geotrichum candidum Dec 1 involved in decolorization of dyes. *Applied and Environmental Microbiology* 65, 1029–1035.

Kobya, K., & Gengec, E. (2012). Decolourization of melanoidins by a electrocoagulation process using aluminium electrodes. *Environmental Technology* 33(21), 2429–2438.

Krishnamoorthy, S., Manickam, P., & Muthukaruppan, V. (2019). Evaluation of distillery wastewater treatability in a customized photobioreactor using blue-green microalgae – Laboratory and outdoor study. *Journal of Environmental Management* 234, 412–423.

Krishnamoorthy, S., Premalatha, M., & Vijayasekaran, M. (2017). Characterization of distillery wastewater – An approach to retrofit existing effluent treatment plant operation with phycoremediation. *Journal of Cleaner Production* 148, 735–750.

Kumar, A., Saroj, D.P., Tare, V., & Bose, P. (2006). Treatment of distillery spent-wash by ozonation and biodegradation: significance of pH reduction and inorganic carbon removal before ozonation. *Ecotoxicology and Environmental Safety* 78(9), 994–1004.

Kumar, G.S., Gupta, S.K., & Singh, G. (2007). Biodegradation of distillery spent wash in anaerobic hybrid reactor. *Water Research* 41(4), 721–730.

Kumar, P., & Chandra, R. (2004). Detoxification of distillery effluent through Bacillus thuringiensis (MTCC 4714) enhanced phytoremediation potential of *Spirodela polyrrhiza* (L.) Schliden. *Bulletin of Environmental Contamination and Toxicology* 73, 903–910.

Kumar, S., & Gopal, K. (2001). Impact of distillery effluent on physiological consequences in the Freshwater teleost. *Bulletin of Environmental Contamination and Toxicology* 66(5), 0617–0622.

Kumar, V. (2018a). Mechanism of microbial heavy metal accumulation from polluted environment and bioremediation. In: Sharma, D., Saharan, B.S. (Eds), *Microbial fuel factories*. CRC Press: Boca Raton.

Kumar, V. (2018b). *Study of bacterial communities in two-step treatment of post methanated distillery effluent by bacteria and constructed wetland plant treatment system*. PhD Thesis, Babasaheb Bhimrao Ambedkar University, Lucknow, India.

Kumar, V. (2021). Phytoremediation of distillery effluent: Current progress, challenges, and future opportunities. In: Saxena, G., Kumar, V., Shah, M.P. (Eds), *Bioremediation for environmental sustainability: Toxicity, mechanisms of contaminants degradation, detoxification and challenges*. Elsevier.

Kumar, V., & Chandra, R. (2018a). Characterisation of manganese peroxidase and laccase producing bacteria capable for degradation of sucrose glutamic acid-maillard products at different nutritional and environmental conditions. *World Journal of Microbiology and Biotechnology* 34, 32.

Kumar, V., & Chandra, R. (2018b) Bacterial assisted phytoremediation of industrial waste pollutants and eco-restoration. In: Chandra, R., Dubey, N.K., Kumar, V. (Eds), *Phytoremediation of environmental pollutants*. CRC Press: Boca Raton.

Kumar, V., & Chandra, R. (2020a). Bioremediation of melanoidins containing distillery waste for environmental safety. In: Saxena, G., Bharagava, R.N. (Eds), *Bioremediation of industrial waste for environmental safety: Microbes and methods for industrial waste management*. Springer Nature: Singapore.

Kumar, V., & Chandra, R. (2020b). Metagenomics analysis of rhizospheric bacterial communities of *Saccharum arundinaceum* growing on organometallic sludge of sugarcane molasses-based distillery. *3 Biotech* 10(7), 1–18.

Kumar, V., & Chandra, R. (2020c). Bacterial-assisted phytoextraction mechanism of heavy metals by native hyperaccumulator plants from distillery waste–contaminated site for eco-restoration. In: Chandra, R., Sobti, R. (Eds), *Microbes for sustainable development and bioremediation*. CRC Press: Boca Raton.

Kumar, A., & Chandra, R. (2020d). Ligninolytic enzymes and its mechanisms for degradation of lignocellulosic waste in environment. *Heliyon* 6(2), e03170.

Kumar, V., Chandra, R., Thakur, I.S., Saxena, G., & Shah, M.P. (2020). Recent advances in physicochemical and biological treatment approached for distillery wastewater. In: Shah, M.P., Banerjee, A. (Eds), *Combine application of physico-chemical and microbiological process for industrial effluent*. Springer Nature: Singapore. https://doi.org/0.1007/978-981-15-0497-6_14

Kumar, V., & Chopra, A.K. (2012). Fertigation effect of distillery effluent on agronomical practices of *Trigonella foenum-graecum* L. (Fenugreek). *Environmental Monitoring and Assessment* 184(3), 1207–1219.

Kumar, V., Kaushal, A., Singh, K., & Shah, M.P. (2021c). Phytoaugmentation technology for phytoremediation of environmental pollutants: Opportunities, challenges and future prospects. In: Kumar, V., Saxena, G., Shah, M.P. (Eds), *Bioremediation for environmental sustainability: Approaches to tackle pollution for cleaner and greener society*. Elsevier

Kumar, V., Shahi, S.K., & Singh, S. (2018). Bioremediation: An eco-sustainable approach for restoration of contaminated sites. In: Singh, J., Sharma, D., Kumar, G., Sharma, N.R. (Eds), *Microbial bioprospecting for sustainable development*. Springe Nature: Singapore.

Kumar, V., Shahi, S.K., Ferreira, L.F.R., Bilal, M., Biswas, J.K., Bulgariu, L., 2021. Characterization and identification of recalcitrant organic and inorganic pollutants discharged in biomethanated distillery effluent and their phytotoxicity, cytotoxicity and genotoxicity assessment using *Phaseolus aureus* L. and *Allium cepa* L. Environmental Research 201, 111551. https://doi.org/10.1016/j.envres.2021.111551

Kumar, V., & Sharma, D.C. (2019). Distillery effluent: Pollution profile, eco-friendly treatment strategies, challenges, and future prospects. In: Arora, P.K. (Ed) *Microbial metabolism of xenobiotic compounds*. Springer Nature. https://doi.org/10.1007/978-981-13-7462-3_17

Kumar, V., Singh, K., & Shah, M.P. (2021a). Advanced oxidation processes for complex wastewater treatment. In: Shah, M.P. (Ed) *Advanced oxidation processes for complex wastewater treatment*. Elsevier. https://doi.org/10.1016/B978-0-12-821011-6.00001-3

Kumar, V., Singh, K., Shah, M.P., & Kumar, M. (2021b). Phytocapping: An eco-sustainable green technology for cleaner environment. In: Kumar, V., Saxena, G., Shah, M.P. (Eds), *Bioremediation for environmental sustainability: Approaches to tackle pollution for cleaner and greener society*. Elsevier. https://doi.org/10.1016/B978-0-12-820318-7.00022-8

Kumar, V., & Thakur, I.S. (2020a). Biodiesel production from transesterification of *Serratia* sp. ISTD04 lipids using immobilised lipase on biocomposite materials of biomineralized products of carbon dioxide sequestrating bacterium. *Bioresource Technology*, 307, 123193.

Kumar, V., & Thakur, I.S. (2020b). Extraction of lipids and production of biodiesel from secondary tannery sludge by in situ transesterification. *Bioresource Technology Reports* 11(April), 100446.

Kumar, V., Thakur, I.S., & Shah, M.P. (2020a). Bioremediation approaches for treatment of pulp and paper industry wastewater: Recent advances and challenges. In: Shah, M.P. (Ed), *Microbial bioremediation & biodegradation*. Springer Nature: Singapore, pp: 1–48.

Kumar, V., Thakur, I.S., Singh, A.K., & Shah, M.P. (2020b). Application of metagenomics in remediation of contaminated site and environmental restoration. In: Shah, M., Rodriguez-Couto, S., Sengor, S.S. (Eds), *Emerging technologies in environmental bioremediation*. Elsevier. https://doi.org/10.1016/B978-0-12-819860-5.00008-0

Kumar, V., Wati, L., Nigam, P., Banat, I.M., Yadav, B.S., Singh, D., & Marchant, R. (1998). Decolorization and biodegradation of anaerobically digested sugarcane molasses spent wash effluent from biomethanation plants by white-rot fungi. *Process Biochemistry* 33(1), 83–88.

Kumari, K., Ranjan, N., Kumar, S., & Sinha, R.C. (2016). Distillery effluent as a liquid fertilizer: A win-win option for sustainable agriculture. *Environmental Technology (United Kingdom)* 37(3), 381–387.

Kumari, K., Ranjan, N., Ray, S.N.C., Aggarwal, P.K., & Sinha, R.C. (2009). Ecofriendly utilization of diluted distillery effluent on the morphological, physiological and biochemical parameters in the wheat plant "*Triticum aestivum*". *International Journal of Bioved* 20, 1–8.

Kumari, K., Ranjan, N., Sharma, J.P., Aggarwal, P.K., & Sinha, R.C. (2012). Integrated management of diluted distillery effluent and fly ash as a potential biofertilizer: A case study on the vegetative growth and chlorophyll content of the marigold plant, "*Tagetespatula*". *International Journal of Environmental Technology and Management* 15, 275–290.

Kumari, P., Tripathi, B.M., Singh, R.N., Saxena, A.K., & Kaushik, R. (2019). The impact of distillery effluent irrigation on plant-growth-promoting traits and taxonomic composition of bacterial communities in agricultural soil. *Biorxiv*. https://doi.org/10.1101/554709

Langner, E., & Rzeski, W. (2014). Biological properties of melanoidins: A review. *International Journal of Food Properties* 17(2), 344–353.

Latif, M.A., Ghufran, R., Wahid, Z.A., & Ahmad, A. (2011). Integrated application of upflow anaerobic sludge blanket reactor for the treatment of wastewaters. *Water Research* 45(16), 4683–4699.

Lettinga, G. (1995). Anaerobic digestion and wastewater treatment systems. *Antonie Leeuwenhoek* 67, 3–28.

Lettinga, G., van Velseo, A.F.M., Hobma, S.W., & de Zeeuw, W. (1980). Use of the upflow sludge blanket (USB) reactor concept for biological wastewater treatment, especially for anaerobic treatment. *Biotechnology and Bioengineering* 22, 699–734.

Liakos, T.I., & Lazaridis, N.K. (2014). Melanoidins removal from simulated and real wastewaters by coagulation and electro-flotation. *Chemical Engineering Journal* 242, 269–277.

Liakos, T.I., & Lazaridis, N.K. (2016). Melanoidin removal from molasses effluents by adsorption. *Journal of Water Process Engineering* 10, 156–164.

Liang, Z., Wang, Y., Zhou, Y., & Liu, H. (2009a). Coagulation removal of melanoidins from biologically treated molasses wastewater using ferric chloride. *Chemical Engineering Journal* 152(1), 88–94.

Liang, Z., Wang, Y., Zhou, Y., Liu, H., & Wu, Z. (2009b). Variables affecting melanoidins removal from molasses wastewater by coagulation/flocculation. *Separation and Purification Technology* 68(3), 382–389.

Lima, A.A., Montalvao, A.F., Dezotti, M., & Sant'Anna, G.L. (2006). Ozonation of a complex industrial effluent: Oxidation of organic pollutants and removal of toxicity. *Ozone: Science and Engineering* 28(1), 3–8.

Lin, J.C.-T., Liu, Y.-S., & Wang, W.-K. (2020). A full-scale study of high-rate anaerobic bioreactors for whiskey distillery wastewater treatment with size fractionation and metagenomic analysis of granular sludge. *Bioresource Technology* 306, 123032.

Lohchab, R.K., & Saini, J. (2017). Performance of UASB reactor to treat distillery spent wash under various operating conditions. *Annals of Biology* 33(1), 10–15.

López, I., Borzacconi, L., & Passeggi, M. (2017). Anaerobic treatment of sugar cane vinasse : Treatability and real-scale operation. *Journal of Chemical Technology and Biotechnology*

Ma, Y. (2002). Fundamental and applied studies on anaerobic biotechnology for the treatment of high strength cheese processing waste. PhD dissertation, Biological and Agricultural Engineering, UTAH State University, Logan, Utah.

Mabuza, J., Otieno, B., Apollo, S., Matshediso, B., & Ochieng, A. (2017). Investigating the synergy of integrated anaerobic digestion and photodegradation using hybrid photocatalyst for molasses wastewater treatment. *Euro-Mediterranean Journal for Environmental Integration* 2(1), 1–10.

Mahaly, M., Senthilkumar, A.K., Arumugam, S., Kaliyaperumal, C., & Karupannan, N. (2018). Vermicomposting of distillery sludge waste with tea leaf residues. *Sustainable Environment Research* 28(5), 223–227.

Mahar, M.T., Khuhawar, M.Y., Baloch, M.A., & Jahangir, T.M. (2013). Health risk assessment of heavy metals in groundwater, the effect of evaporation ponds of distillery spent wash: A case study of southern Punjab Pakistan. *World Applied Sciences Journal* 28(11), 1748–1756.

Mahgoub, S., Tsioptsias, C., & Samaras, P. (2016). Biodegradation and decolorization of melanoidin solutions by manganese peroxidase yeasts. *Water Science and Technology* 73(10), 2436–2445.

Malik, S.N., Ghosh, P.C., Vaidya, A.N., & Mudliar, S.N. (2019a). Ozone pre-treatment of molasses-based biomethanated distillery wastewater for enhanced bio-composting. *Journal of Environmental Management* 246, 42–50.

Malik, S.N., Khan, S.M., Ghosh, P.C., Vaidya, A.N., Das, S., & Mudliar, S.N. (2019b). Nano catalytic ozonation of biomethanated distillery wastewater for biodegradability enhancement, color and toxicity reduction with biofuel production. *Chemosphere* 230, 449–461.

Malik, S.N., Saratchandra, T., Tembhekar, P.D., Padoley, K.V., Mudliar, S.L., & Mudliar, S.N. (2014). Wet air oxidation induced enhanced biodegradability of distillery effluent. *Journal of Environmental Management* 136, 132–138.

Manach, C., Scalbert, A., Morand, C., Remesy, C., & Jimenez, L. (2004). Polyphenols: Food sources and bioavailability. *American Journal of Clinical Nutrition* 79(5), 727–747.

Mandal, A., Ojha, K., & Ghosh, D.N. (2003). Removal of color from distillery wastewater by different processes. *Indian Journal of Chemistry—Section B* 45, 264–267.

Mani, D., & Kumar, C. (2014). Biotechnological advances in bioremediation of heavy metals contaminated ecosystems: An overview with special reference to phytoremediation. *International Journal of Environmental Science and Technology* 11(3), 843–872.

Manisankar, P., Viswanathan, S., & Rani, C. (2003). Electrochemical treatment of distillery effluent using catalytic anodes. *Green Chemistry* 5(2), 270–274.

Manyuchi, M.M., Mbohwa, C., & Muzenda, E. (2018). Biological treatment of distillery wastewater by application of the vermifiltration technology. *South African Journal of Chemical Engineering* 25, 74–78.

Martin, M.A., Raposo, F., Borja, R., & Martin, A. (2002). Kinetic study of the anaerobic digestion of vinasse pretreated with ozone, ozone plus ultraviolet, and ozone plus ultraviolet light in the presence of titanium dioxide. *Process Biochemistry* 37, 699e706.

Metcalf, E. (2003). *Wastewater engineering*, 4th ed. McGraw Hill Inc: New York, p. 10.

Migo, V.P., Del Rosario, E.J., & Matsumura, M., 1997. Flocculation of melanoidins induced by inorganic ions. *Journal of Fermentation and Bioengineering* 83(3), 287–291.

Miyata, N., Iwahori, K., & Fujita, M. (1998). Manganese-independent and -dependent decolorization of melanoidin by extracellular hydrogen peroxide and peroxidases from *Coriolus hirsutus* pellets. *Journal of Fermentation and Bioengineering* 85(5), 550–553.

Miyata, N., Mori, T., Iwahori, K., & Fujita, M. (2000). Microbial decolorization of melanoidin-containing wastewaters: Combined use of activated sludge and the fungus *Coriolus hirsutus*. *Journal of Bioscience and Bioengineering* 89(2), 145–150.

Mohan, V., Mohanakrishna, S., Ramanaiah, G.S.V., & Sarma, P.N. (2008). Simultaneous biohydrogen production and wastewater treatment in biofilm configured anaerobic periodic discontinuous batch reactor using distillery wastewater. *International Journal of Hydrogen Energy* 33(2), 550–558.

Mohana, S., Desai, C., & Madamwar, D. (2007). Biodegradation and decolourization of anaerobically treated distillery spent wash by a novel bacterial consortium. *Bioresource Technology* 98(2), 333–339.

Mohana, S., Shah, A., Divecha, J., & Madamwar, D. (2008). Xylanase production by *Burkholderia* sp. DMAX strain under solid state fermentation using distillery spent wash. *Bioresource Technology* 99(16), 7553–7564.

Mohanakrishna, G., Venkata Mohan, S., & Sarma, P.N. (2009). Bio-electrochemical treatment of distillery wastewater in microbial fuel cell facilitating decolorization and desalination along with power generation. *Journal of Hazardous Materials* 177(1–3), 487–494.

Momeni, M.M., Kahforoushan, D., Abbasi, F., & Ghanbarian, S. (2018). Using Chitosan/CHPATC as coagulant to remove color and turbidity of industrial wastewater: Optimization through RSM design. *Journal of Environmental Management* 211, 347–355.

Moreira, A.S.P., Nunes, F.M., Domingues, M.R., & Coimbra, M.A. (2012). Coffee melanoidins: Structures, mechanisms of formation and potential health impacts. *Food and Function* 3(9), 903–915.

Mulidzi, A.R. (2010). Winery and distillery wastewater treatment by constructed wetland with shorter retention time. *Water Science and Technology* 61(10), 2611–2615.

Murata, M., Terasawa, N., & Homma, S. (1992). Screening of microorganisms to decolorize a model melanoidin and the chemical properties of a microbially treated melanoidin. *Bioscience, Biotechnology, and Biochemistry* 56(8), 1182–1187.

Murphy, C., Hawes, P., & Cooper, D.J. (2009). The application of wetland technology for copper removal from distillery wastewater: A case study. *Water Science and Technology* 60(11), 2759–2766.

Musee, N., Trerise, M.A., & Lorenzen, L. (2007). Post-treatment of Distillery Wastewater after UASB using Aerobic Techniques. *South African Journal of Enology and Viticulture* 28(1), 55–55.

Nagarajan, S., & Ranade, V.V. (2020). Pre-treatment of distillery spent wash (vinasse) with vortex based cavitation and its influence on biogas generation. *Bioresource Technology Reports* 11, 100480.

Naina Mohamed, S., Thota Karunakaran, R., & Manickam, M. (2018). Enhancement of bioelectricity generation from treatment of distillery wastewater using microbial fuel cell. *Environmental Progress and Sustainable Energy* 37(2), 663–668.

Nandy, T., Shastry, S., & Kaul, S.N. (2002). Wastewater management in a cane molasses distillery involving bioresource recovery. *Journal of Environmental Management* 65(1), 25–38.

Narain, K., Bhat, M.M., & Yunus, M. (2012). Impact of distillery effluent on germination and seedling growth of *Pisum sativum* L. *Universal Journal of Environmental Research and Technology* 2(4), 269–272.

Narain, K., Yazdani, T., Muzamil Bhat, M., & Yunus, M. (2012). Effect on physico-chemical and structural properties of soil amended with distillery effluent and ameliorated by cropping two cereal plant spp. *Environmental Earth Sciences* 66(3), 977–984.

Nataraj, S.K., Hosamani, K.M., & Aminabhavi, T.M. (2006). Distillery wastewater treatment by the membrane based nanofiltration and reverse osmosis. *Water Research* 40, 2349–2356.

Nayak, J.K., Amit, & Ghosh, U.K. (2018). An innovative mixotrophic approach of distillery spent wash with sewage wastewater for biodegradation and bioelectricity generation using microbial fuel cell. *Journal of Water Process Engineering* 23, 306–313.

Nguyen, T., Fan, L., & Roddick, F. (2010). Removal of melanoidins from an industrial wastewater. *Ozwater* 10, 2, 1–4. http://www.awa.asn.au/uploadedFiles/Remove melanoidinsfrom Industrial wastewater.pdf

Ogunwole, E., Kunle-Alabi, O.T., Akindele, O.O., & Raji, Y. (2020). *Saccharum officinarum* molasses adversely alters reproductive functions in male wistar rats. *Toxicology Reports* 7, 345–352.

Ohmomo, S., Aoshima, I., Tozawa, Y., Sakurada, N., & Ueda, K. (1985). Purification and some properties of melanoidin decolorizing enzymes, P-III and P-IV, from mycelia of *Coriolus versicolor* Ps4a. *Agricultural and Biological Chemistry* 49(7), 2047–2053.

Ohmomo, S., Kainuma, M., Kamimura, K., Sirianuntapiboon, S., Aoshima, I., & Atthasampunna, P. (1988). Adsorption of melanoidin to the mycelia of *Aspergillus oryzae* Y-2-32. *Agricultural and Biological Chemistry* 52(2), 381–386.

Ohmomo, S., Kaneko, Y., Sirianuntapiboon, S., Somchai, P., Atthasampunna, P., & Nakamura, I. (1987). Decolorization of molasses waste water by a thermophilic strain, *Aspergillus fumigatus* G-2-6 Sadahiro. *Agricultural and Biological Chemistry* 51(12), 3339–3346.

Olguın, E.J., Sanchez-Galvan, G., Gonzalez-Portela, R.E., & Lopez-Vela, M. (2008). Constructed wetland mesocosms for the treatment of diluted sugarcane molasses stillage from ethanol production using Pontederia sagittata. *Water Research* 42(14), 3659–3666.

Olguın, E.J., Sanchez-Galvan, G., Perez-Perez, T. (2007). Assessment of the phytoremediation potential of Salvinia minima Baker compared to Spirodela polyrrhiza in highstrength organic wastewater. *Water, Air, & Soil Pollution* 181 (1–4), 135–147.

Omar, A.A., Mahgoub, S., Salama, A., Likotrafiti, E., Rhoades, J., & Christakis, C., Samaras, P. (2021). Evaluation of *Lactobacillus kefiri* and manganese peroxidase-producing bacteria for decolorization of melanoidins and reduction of chemical oxygen demand. *Water and Environmental Journal* 35(2), 704–714.

Oosterkamp, M.J., Méndez-García, C., Kim, C.-H., Bauer, S., Ibáñez, A.B., Zimmerman, S., Hong, P.-Y., Cann, I.K., & Mackie, R.I. (2016). Lignocellulose-derived thin stillage composition and efficient biological treatment with a high-rate hybrid anaerobic bioreactor system. *Biotechnology for Biofuels* 9, 120. https://doi.org/10.1186/s13068-016-0532-z

Padoley, K.V., Saharan, V.K., Mudliar, S.N., Pandey, R., & Pandit, A.B. (2012a). Cavitationally induced biodegradability enhancement of a distillery wastewater. *Journal of Hazardous Materials* 219–220, 69–74.

Padoley, K.V., Tembhekar, P.D., Saratchandra, T., Pandit, A.B., Pandey, R.A., & Mudliar, S.N. (2012b). Wet air oxidation as a pretreatment option for selective biodegradability enhancement and biogas generation potential from complex effluent. *Bioresource Technology* 120, 157–164.

Pandey, S.N., Nautiyal, B.D., & Sharma, C.P. (2008). Pollution level in distillery effluent and its phytotoxic effect on seed germination and early growth of maize and rice. *Journal of Environmental Biology* 29(2), 267–270.

Pant, D., & Adholeya, A. (2007a). Enhanced production of ligninolytic enzymes and decolorization of molasses distillery wastewater by fungi under solid state fermentation. *Biodegradation* 18(5), 647–659.

Pant, D., & Adholeya, A. (2007b). Identification, ligninolytic enzyme activity and decolorization potential of two fungi isolated from a distillery effluent contaminated site. *Water, Air, and Soil Pollution* 183(1–4), 165–176.

Pant, D., & Adholeya, A. (2009a). Concentration of fungal ligninolytic enzymes by ultrafiltration and their use in distillery effluent decolorization. *World Journal of Microbiology and Biotechnology* 25(10), 1793–1800.

Pant, D., & Adholeya, A. (2009b). Nitrogen removal from biomethanated spentwash using hydroponic treatment followed by fungal decolorization. *Environmental Engineering Science* 26(3), 559–565.

Pathak, H., Joshi, H.C., Chaudhary, A., Chaudhary, R., Kalra, N., & Dwiwedi, M.K. (1999). Soil amendment with distillery effluent for wheat and rice cultivation. *Water, Air, and Soil Pollution* 113(1–4), 133–140.

Pazouki, M., Najafpour, G., & Hosseini, M.R. (2008). Kinetic models of cell growth, substrate utilization and bio-decolorization of distillery wastewater by *Aspergillus fumigatus* UB260. *African Journal of Biotechnology* 7(9), 1369–1376.

Pazouki, M., Najafpour, G., & Hosseini, M.R. (2017). Kinetic models of cell growth, substrate utilization and bio-decolorization of distillery wastewater by *Aspergillus fumigatus*. *International Journal of Histology and Cytology* 4(7), 337–344.

Prajapati, A.K., & Chaudhari, P.K. (2015). Physicochemical treatment of distillery wastewater – A review. In: Chemical engineering communications.

Prajapati, A.K., Chaudhari, P.K., Pal, D., Chandrakar, A., & Choudhary, R. (2016). Electrocoagulation treatment of rice grain based distillery effluent using copper electrode. *Journal of Water Process Engineering* 11(June), 1–7.

Prasad, R.K., & Srivastava, S.N. (2009). Electrochemical degradation of distillery spent wash using catalytic anode : Factorial design of experiments. *Chemical Engineering Journal* 146(1), 22–29.

Prodanovic, J.M., & Vasic, V.M. (2013). Application of membrane processes for distillery wastewater purification—A review. *Desalination and Water Treatment* 51(16–18), 3325–3334.

Quan, X., Tao, K., Mei, Y., & Jiang, X. (2014). Power generation from cassava alcohol wastewater: Effects of pretreatment and anode aeration. *Bioprocess and Biosystems Engineering* 37(11), 2325–2332.

Rafigh, S.M., & Rahimpour Soleymani, A. (2019). Melanoidin removal from molasses wastewater using graphene oxide nanosheets. *Separation Science and Technology* 55(13), 2281–2293.

Raghukumar, C., Mohandass, C., Kamat, S., & Shailaja, M.S. (2004). Simultaneous detoxiWcation and decolorization of molasses spent wash by the immobilized white rot fungus Flavadon Xavus isolated from marine habitat. *Enzyme and Microbial Technology* 35, 197–202.

Raghukumar, C., & Rivonkar, G. (2001). Decolorization of molasses spent wash by the white-rot fungus Flavodon flavus, isolated from a marine habitat. *Applied Microbiology and Biotechnology* 55, 510–514.

Rai, U., Muthukrishnan, M., & Guha, B.K. (2008). Tertiary treatment of distillery wastewater by nanofiltration. *Desalination* 230(1–3), 70–78.

Rakkiyappan, P., Thangavelu, S., Malathi, R., Radhamani, R. (2001). Effect of biocompost and enriched pressmud on sugarcane yield and quality. *Sugar Tech* 3(3), 92–96.

Ramakritinan, C.M., Kumaraguru, A.K., & Balasubramanian, M.P. (2005). Impact of distillery effluent on carbohydrate metabolism of freshwater fish, Cyprinus carpio. *Ecotoxicology* 14(7), 693–707.

Ramana, S., Biswas, A.K., Kundu, S., Saha, J.K., & Yadava, R.B.R. (2002). Effect of distillery effluent on seed germination in some vegetable crops. *Bioresource Technology* 82, 273–275.

Rani, K., Sridevi, V., Rao, R.S.V., Kumar, K.V., & Harsha, N. (2013). Biological treatment of distillery waste water – An overview. *International Journal of General Engineering and Technology (IJGET)* 2(4), 15–24.

Rao, S. B., 1972. A low cost waste treatment method for disposal of distillery waste (spent wash). *Water Research* 6, 1275–1282.

Rath, P., Pradhan, G., & Mishra, M.K. (2010). Effect of sugar factory distillery spent wash (DSW) on the growth pattern of sugarcane (*Saccharum officinarum*) crop. *Journal of Phytology* 2(5), 33–39.

Ravikumar, R. (2015). Effect of transport phenomena of *Cladosporium cladosporioides* on decolorization and chemical oxygen demand of distillery spent wash. *International Journal of Environmental Science and Technology* 12(5), 1581–1590.

Ravikumar, R., & Karthik, V. (2015). Effective utilization and conversion of spent distillery liquid to valuable products using an intensified technology of two-stage biological sequestration. *Chemical and Biochemical Engineering Quarterly* 29(4), 599–608.

Ravikumar, R., Vasanthi, N.S., & Saravanan, K. (2011). Single factorial experimental design for decolorizing anaerobically treated distillery spent wash using *Cladosporium cladosporioides*. *International Journal of Environmental Science and Technology* 8(1), 97–106.

Ravikumar, R., Vasanthi, N.S., & Saravanan, K. (2013). Biodegradation and decolorization of distillery spent wash with product release by a novel strain *Cladosporium cladosporioides:* Optimization and biokinetics. *Chemical and Biochemical Engineering Quarterly* 27(3), 373–383.

Reis, C.E.R., Bento, H.B.S., Alves, T.M., Carvalho, A.K.F., & De Castro, H.F. (2019). Vinasse treatment within the sugarcane-ethanol industry using ozone combined with anaerobic and aerobic microbial processes. *Environments* 6(1), 5.

Reis, C.E.R., & Hu, B. (2017). Vinasse from sugarcane ethanol production: Better treatment or better utilization? *Frontiers in Energy Research* 5(APR), 1–7.

Rioja, R., Garcia, M.T., Pena, M., & Gonzalez, G. (2008). Biological decolourisation of wastewater from molasses fermentation by *Trametes versicolor* in an airlift reactor. *Journal of Environmental Science and Health, Part A: Toxic/Hazardous Substances and Environmental Engineering* 43(7), 772–778.

Rizvi, S., Goswami, L., & Gupta, S.K. (2020). A holistic approach for melanoidin removal via Fe-impregnated activated carbon prepared from *Mangifera indica* leaves biomass. *Bioresource Technology Reports* 12,100591.

Robert, S.A., Wildner, G.F., Grass, G., Weichse, A., Amrus, A., Rensing, C., & Montfort, W.R.A. (2003). Regulatory copper ion lies near the T1 copeer site in the muclticopper oxidase CueO. *Journal of Biological Chemistry* 278:31958–31963.

Rocha, M.H., Lora, E.E.S., & Venturini, O.J. (2007). Life cycle analysis of different alternatives for the treatment and disposal of ethanol vinasse. *Sugar Industry* 133(2), 88–93.

Rodrigues Reis, C. E., Valle, G. F., Bento, H.B.S., Carvalho, A.K.F., Alves, T.M., & de Castro, H.F. (2020). Sugarcane by-products within the biodiesel production chain: Vinasse and molasses as feedstock for oleaginous fungi and conversion to ethyl esters. *Fuel* 277, 118064.

Rodríguez, S., Fernández, M., Bermúdez, R.C., & Morris, H. (2003). Tratamiento de efluentes industriales coloreados com *Pleurotus* spp. *Revista Iberoamericana de Micología* 20, 164–168.

Rodriguez-caballero, A., Ramond, J., Welz, P.J., Cowan, D.A., Odlare, M., & Burton, S.G. (2012). Treatment of high ethanol concentration wastewater by biological sand filters : Enhanced COD removal and bacterial community dynamics. *Journal of Environmental Management* 109, 54–60.

Romanholo Ferreira, L.F., Aguiar, M., Pompeu, G., Messias, T.G., & Monteiro, R.R. (2010). Selection of vinasse degrading microorganisms. *World Journal of Microbiology & Biotechnology* 26, 1613–1621.

Ruhi, R.A., Saha, A.K., Rahman, S. M. A., Mohanta, M. K., Sarker, S. R., Nasrin, T., & Haque, M.F. (2017). Decolourization of synthetic melanoidin by bacteria isolated from sugar mill effluent. *University Journal of Zoology, Rajshahi University* 36, 12–21.

Rulli, M.M., Villegas, L.B., & Colin, V.L. (2020). Treatment of sugarcane vinasse using an autochthonous fungus from the northwest of Argentina and its potential application in fertigation practices. *Journal of Environmental Chemical Engineering* 8(5), 104371.

Sa, G., Gonza, R.E., Lo, M., Olgui, E.J., & Haya, E. (2008). Constructed wetland mesocosms for the treatment of diluted sugarcane molasses stillage from ethanol production using *Pontederia sagittata*. *Water Research* 42, 3659–3666.

Saini, J.K., & Lohchab, R.K. (2017). Performance of UASB reactor to treat distillery spent wash under various operating conditions. *Annals of Biology* 33 (1), 10–15.

Sales, D., Valcárcel, M.J., Pérez, L., & Ossa, E. (1987). Activated sludge treatment of wine-distillery wastewaters. *Journal of Chemical Technology & Biotechnology* 40(2), 85–99. https://doi.org/10.1002/jctb.280400202

Sánchez-Galván, G., & Bolaños-Santiago, Y. (2018). Phytofiltration of anaerobically digested sugarcane ethanol stillage using a macrophyte with high potential for biofuel production. *International Journal of Phytoremediation* 20(8), 805–812.

Sánchez-Galván, G., Torres-Quintanilla, E., Sayago, J., & Olguín, E.J. (2015). Color removal from anaerobically digested sugar cane stillage by biomass from invasive macrophytes. *Water, Air, and Soil Pollution* 226, 110.

Saner, A.B., Mungray, A.K., & Mistry, N.J. (2014). Treatment of distillery wastewater in an upflow anaerobic sludge blanket (UASB) reactor. *Desalination and Water Treatment* 57(10), 4328–4344.

Sangave, P.C., Gogate, P.R., & Pandit, A.B. (2007a). Combination of ozonation with conventional aerobic oxidation for distillery wastewater treatment. *Chemosphere* 68(1), 32–41.

Sangave, P.C., Gogate, P.R., & Pandit, A.B. (2007b). Ultrasound and ozone assisted biological degradation of thermally pretreated and anaerobically pretreated distillery wastewater. *Chemosphere* 68, 42–50.

Sangave, P.C., & Pandit, A.B. (2004). Ultrasound pre-treatment for enhanced biodegradability of the distillery wastewater. *Ultrasonics Sonochemistry* 11 (3–4), 197–203.

Sankaran, K., Pisharody, L., Suriya Narayanan, G., & Premalatha, M. (2015). Bacterial assisted treatment of anaerobically digested distillery wastewater. *RSC Advances* 5(87), 70977–70984.

Sankaran, K., & Premalatha, M. (2018). Nutrients uptake from anaerobically digested distillery wastewater by *Spirulina* sp. under xenon lamp illumination. *Journal of Water Process Engineering* 25, 295–300.

Sankaran, K., Premalatha, M., Vijayasekaran, M., & Somasundaram, V.T. (2014). DEPHY project: Distillery wastewater treatment through anaerobic digestion and phycoremediation – A green industrial approach. *Renewable and Sustainable Energy Reviews* 37, 634–643.

Sankaran, S., Khanal, S.K., Jasti, N., Jin, B., Pometto, A.L., & Van Leeuwen, J.H. (2010). Use of filamentous fungi for wastewater treatment and production of high value fungal byproducts: A review. *Critical Reviews in Environmental Science and Technology* 40(5), 400–449. https://doi.org/10.1080/10643380802278943

Santal, A.R., Singh, N.P., & Saharan, B.S. (2011). Biodegradation and detoxification of melanoidin from distillery effluent using an aerobic bacterial strain SAG_5 of *Alcaligenes faecalis*. *Journal of Hazardous Materials* 193, 319–324.

Santal, A.R., Singh, N.P., & Saharan, B.S. (2016). A novel application of *Paracoccus pantotrophus* for the decolorization of melanoidins from distillery effluent under static conditions. *Journal of Environmental Management* 169, 78–83.

Sarat Chandra, T., Malik, S.N., Suvidha, G., Padmere, M.L., Shanmugam, P., & Mudliar, S.N. (2014). Wet air oxidation pretreatment of biomethanated distillery effluent: Mapping pretreatment efficiency in terms color, toxicity reduction and biogas generation. *Bioresource Technology* 158(February), 135–140.

Satyawali, Y., & Balakrishnan, M. (2007). Removal of color from biomethanated distillery spentwash by treatment with activated carbons. *Bioresource Technology* 98(14), 2629–2635.

Satyawali, Y., & Balakrishnan, M. (2008). Wastewater treatment in molasses-based alcohol distilleries for COD and color removal: A review. *Journal of Environmental Management* 86(3), 481–497. https://doi.org/10.1016/j.jenvman.2006.12.024

Sandermann, H., 1994. Higher plant metabolism of xenobiotics: the 'green liver' concept. *Pharmacogenetics* 4(5), 225–241.

Sharif, M., Shahzad, M.A., Rehman, S., Khan, S., Ali, R., Khan, M.L., & Khan, K. (2012). Nutritional evaluation of distillery sludge and its effect as a substitute of canola meal on performance of broiler chickens. *Journal of Animal Science* 25(3), 401–409.

Sharma, J., & Singh, R. (2001). Effect of nutrient supplementation on anaerobic sludge development and activity for treating distillery effluent. *Bioresource Technology* 79, 203–206.

Sharma, K.P., Singh, P.K., Kumar, S., Sharma, S., & Kumar, R. (2011). Tolerance of some hardy plant species to biomethanated spent wash of distilleries. *Indian Journal of Biotechnology* 10(1), 97–112.

Sharma, S., Sharma, K., Yadav, N., Ojha, K., Sharma, S., & Sharma, K.P. (2011). Efficacy of distillery soil leachate on reproductive health of Swiss albino male mice (*Mus musculus* L.). *Pharmacologyonline* 2, 748–754.

Shinde, P.A., Ukarde, T.M., Pandey, P.H., & Pawar, H.S. (2020). Distillery spent wash : An emerging chemical pool for next generation sustainable distilleries. *Journal of Water Process Engineering* 36, 101353.

Sievers, M. (2011). Advanced oxidation processes. *Treatise on Water Science* 4, 377–408.

Singh, D., & Nigam, P. (1995). Treatment and disposal of distillery effluents in India. In: Moo-Young M., Anderson W.A., Chakrabarty A.M. (eds) *Environmental Biotechnology*. Springer, Dordrecht. https://doi.org/10.1007/978-94-017-1435-8_64.

Singh, J., Kaur, A., & Vig, A.P. (2014). Bioremediation of distillery sludge into soil-enriching material through vermicomposting with the help of *Eisenia fetida*. *Applied Biochemistry and Biotechnology* 174(4), 1403–1419.

Singh, N., Basu, S., Vankelecom, I.F.J., & Balakrishnan, M. (2015). Covalently immobilized laccase for decolourization of glucose-glycine maillard products as colourant of distillery wastewater. *Applied Biochemistry and Biotechnology* 177(1), 76–89.

Singh, N., Petrinic, I., Hélix-nielsen, C., Basu, S., & Balakrishnan, M. (2019). Influence of forward osmosis (FO) membrane properties on dewatering of molasses distillery wastewater. *Journal of Water Process Engineering* 32(August), 100921.

Singh, N.K., Pandey, G.C., Rai, U.N., Tripathi, R.D., Singh, H.B., Gupta, D.K. (2005). Metal accumulation and ecophysiological effects of distillery effluent on *Potamogeton pectinatus* L. *The Bulletin of Environmental Contamination and Toxicology* 74, 857–863.

Singh, P.K., & Sharma, K.P. (2012). Biomethanated distillery wastae utilization by wetland plant: A New approach. *World Journal of Agricultural Sciences* 8(1), 96–103.

Singh, P.K., Sharma, K.P., Sharma, S., Swami, R.C., & Sharma, S. (2010). Polishing of biomethanated spent wash (primary treated) in constructed wetland: A bench scale study. *Indian Journal of Biotechnology* 9(3), 313–318.

Singh, S., & Dikshit, A.K. (2012). Decolourization of polyaluminium chloride and fungal sequencing batch aerobic reactor treated molasses spentwash by ozone. *American Journal of Environmental Engineering* 2(3), 45–48.

Sinha, S.K., Jha, C.K., Kumar, V., Kumari, G., & Alam, M. (2014). Integrated effect of bio-methanated distillery effluent and bio-compost on soil properties, juice quality and yield of sugarcane in entisol. *Sugar Technology* 16(1), 75–79.

Sirianuntapiboon, S., Phothilangka, P., & Ohmomo, S. (2004). Decolorization of molasses wastewater by a strain No. BP103 of acetogenic bacteria. *Bioresource Technology* 92(1), 31–39.

Sirianuntapiboon, S., Sihanonth, P., Somchai, P., Atthasampunna, P., & Hayashida, S. (1995). An absorption mechanism for the decolorization of melanoidin by *Rhizoctonia* sp. D-90. *Bioscience, Biotechnology, and Biochemistry* 59(7), 1185–1189.

Sirianuntapiboon, S., Zohsalam, P., & Ohmomo, S. (2004). Decolorization of molasses wastewater by *Citeromyces* sp. WR-43-6. *Process Biochemistry* 39(8), 917–924.

Solovchenko, A., Pogosyan, S., Chivkunova, O., Selyakh, I., Semenova, L., Voronova, E., Scherbakov, P., Konyukhov, I., Chekanov, K., Kirpichnikov, M., & Lobakova, E. (2014). Phycoremediation of alcohol distillery wastewater with a novel *Chlorella sorokiniana* strain cultivated in a photobioreactor monitored on-line via chlorophyll fluorescence. *Algal Research* 6(Part B), 234–241.

Sonawane, J.M., Marsili, E., & Ghosh, P.C. (2014). Treatment of domestic and distillery wastewater in high surface microbial fuel cells. *International Journal of Hydrogen Energy* 39(36), 21819–21827.

Srivastava, P.C., Singh, A.P., Kumar, S., Ramachandran, V., Shrivastava, M., & D'souza, S.F. (2009). Efficacy of phosphorus enriched post-methanation bio-sludge from molasses based distillery as P source to rice and wheat crops grown in a mollisol: I. Laboratory and greenhouse evaluation with 32P-labeled sources. *Geoderma* 149(3–4), 312–317.

Suthar, S., & Singh, S. (2008). Feasibility of vermicomposting in biostabilization of sludge from a distillery industry. *Science of the Total Environment* 394(2–3), 237–243.

Takle, S.P., Naik, S.D., Khore, S.K., Ohwal, S.A., Bhujbal, N.M., Landge, S.L., Kale, B.B., & Sonawane, R.S. (2018). Photodegradation of spent wash, a sugar industry waste, using vanadium-doped TiO$_2$ nanoparticles. *RSC Advances* 8(36), 20394–20405.

Tapia-Tussell, R., Pérez-Brito, D., Torres-Calzada, C., Cortés-Velázquez, A., Alzate-Gaviria, L., Chablé-Villacís, R., & Solís-Pereira, S. (2015). Laccase gene expression and vinasse biodegradation by *Trametes hirsuta strain* Bm-2. *Molecules* 20(8), 15147–15157.

Taskin, M., Ortucu, S., Unver, Y., Tasar, O.C., Ozdemir, M., & Kaymak, H.C. (2016). Invertase production and molasses decolourization by cold-adapted filamentous fungus *cladosporium herbarum* ER-25 in nonsterile molasses medium. *Process Safety and Environmental Protection* 103, 136–143.

Tembhekar, P.D., Padoley, K.V., Mudliar, S.L., & Mudliar, S.N. (2015). Kinetics of wet air oxidation pretreatment and biodegradability enhancement of a complex industrial wastewater. *Journal of Environmental Chemical Engineering* 3, 339–348.

Thakkar, A.P., Dhamankar, V.S., & Kapadnis, B.P. (2006). Biocatalytic decolorization of molasses by *phaenerochete chrysosporium*. *Bioresource Technology* 97, 1377–1381.

Thakur, C., Srivastava, V.C., & Mall, I.D. (2009). Electrochemical treatment of a distillery wastewater: Parametric and residue disposal study. *Chemical Engineering Journal* 148, 496–505.

Thiyagu, R., & Sivarajan, P. (2018). Isolation and characterization of novel bacterial strain present in a lab scale hybrid UASB reactor treating distillery spent wash. 3330(May).

Tiwari, S., Gaur, R., & Singh, R. (2012). Decolorization of a recalcitrant organic compound (melanoidin) by a novel thermotolerant yeast, *Candida tropicalis* RG-9. *BMC Biotechnology* 12, 13. https://doi.org/10.1186/1472-6750-12-30

Tondee, T., & Sirianutapiboon, S. (2006). Screening of melanoidin decolorization activity in yeast strain. *International Conference on Environment* 99, 5511–5519.

Tondee, T., & Sirianuntapiboon, S. (2008). Decolorization of molasses wastewater by *Lactobacillus plantarum* No. PV71-1861. *Bioresource Technology* 99(14), 6258–6265.

Tondee, T., Sirianuntapiboon, S., & Ohmomo, S. (2008). Decolorization of molasses wastewater by yeast strain, *Issatchenkia orientalis* No. SF9-246. *Bioresource Technology* 99(13), 5511–5519.

Tripathi, D.M., Tripathi, S., & Tripathi, B.D. (2011). Implications of secondary treated distillery effluent irrigation on soil cellulase and urease activities. *Journal of Environmental Protection* 2(5), 655–661.

Tripathy, B.K., Johnson, I., & Kumar, M. (2020). Melanoidin removal in multi-oxidant supplemented microwave system: Optimization of operating conditions using response surface methodology and cost estimation. *Journal of Water Process Engineering* 33, 101008.

Trivedy, R.K., & Nakate, S.S. (2000). Treatment of diluted distillery waste by constructed wetlands. Indian. *The Journal of Environmental Protection* 20(10), 749–753.

USEPA. (2002). The environment protection rules, 3A, Schedule-II, III. U.S. Environmental Protection Agency, Office of research and development, Cincinnati, OH.

Valderrama, L.T., Del Campo, C.M., Rodriguez, C.M., de-Bashan, L.E., & Bashan, Y. (2002). Treat- ment of recalcitrant wastewater from ethanol and citric acid production using the microalga Chlorella vulgaris and the macrophyte *Lemna minuscule*. *Water Research* 36, 4185–4192.

Veeresh, S.J., & Narayana, J. (2013). Earthworm density, biomass and vermicompost recovery during agroindustrial waste treatment. *International Journal of Pharma and Bio Sciences* 4, 1274–1280.

Velásquez-Riaño, M., Carvajal-Arias, C.E., Rojas-Prieto, N.L., Ausecha-García, S.A., Vera-Díaz, M.Á., Meneses-Sánchez, J.S., & Villa-Restrepo, A.F. (2018). Evaluation of a mixed simultaneous vinasse degradation treatment using *Komagataeibacter kakiaceti* GM5 and *Trametes versicolor* DSM 3086. *Ecotoxicology and Environmental Safety* 164, 425–433.

Venkata Mohan, S., Mohanakrishna, G., Ramanaiah, S.V., & Sarma, P.N. (2008). Simultaneous biohydrogen production and wastewater treatment in biofilm configured anaerobic periodic discontinuous batch reactor using distillery wastewater. *International Journal of Hydrogen Energy* 33(2), 550–558.

Verma, A.K., Raghukumar, C., & Naik, C.G. (2011). A novel hybrid technology for remediation of molasses-based raw effluents. *Bioresource Technology* 102(3), 2411–2418.

Vivekanandam, S., Muthunarayanan, V., Muniraj, S., Rhyman, L., Alswaidan, I.A., & Ramasami, P. (2019). Ingenious bioorganic adsorbents for the removal of distillery based pigment-melanoidin: Preparation and adsorption mechanism. *Journal of Macromolecular Science, Part A: Pure and Applied Chemistry* 56(1), 52–62.

Vlissidis, A., & Zouboulis, A.I. (1993). Thermophilic anaerobic digestion of alcohol distillery wastewaters. *Bioresource Technology* 43, 131–140.

Wagh, M.P., & Nemade, P.D. (2018). Biodegradation of anaerobically treated distillery spent wash by aspergillus species from a distillery effluent contaminated site. *Desalination and Water Treatment* 104, 234–240.

Wang, H.-Y., Qian, H., & Yao, W.-R., 2011. Melanoidins produced by the Maillard reaction: Structure and biological activity. *Food Chemistry* 128, 573–584.

Walter, A., Dolzan, P., Quilodrán, O., de Oliveira, J.G., da Silva, C., & Piacente, F., et al. (2011). Sustainability assessment of bio-ethanol production in Brazil considering land use change, GHG emissions and socio-economic aspects. *Energy Policy* 39, 5703–5716.

Watanabe, Y., Sugi, R., Tanaka, Y., & Hayashida, S. (1982). Enzymatic decolorization of melanoidin by *Coriolus* sp. No. 20. *Agricultural and Biological Chemistry* 46(6), 1623–1630.

Wilk, M., & Krzywonos, M. (2020). Distillery wastewater decolorization by *Lactobacillus plantarum MiLAB393*. *Archives of Environmental Protection* 46(1), 76–84.

Wilk, M., Krzywonos, M., Borowiak, D., & Seruga, P. (2019). Decolourization of sugar beet molasses vinasse by lactic acid bacteria – The effect of yeast extract dosage. *Polish Journal of Environmental Studies* 28(1), 385–392.

Wolmarans, B., & de Villiers, G.H. (2002). Start-up of a UASB effluent treatment plant on distillery wastewater. *Water SA* 28, 63–68.

Wong, D.W. (2009). Structure and action mechanism of ligninolytic enzymes. *Applied Biochemistry and Biotechnology* 157(2), 174–209.

Yadav, S., & Chandra, R. (2011). Heavy metals accumulation and ecophysiological effect on *Typha angustifolia* L. and *Cyperus esculentus* L. growing in distillery and tannery effluent polluted natural wetland site, Unnao, India. *Environmental Earth Sciences* 62(6), 1235–1243.

Yadav, S., & Chandra, R. (2012). Biodegradation of organic compounds of molasses melanoidin (MM) from biomethanated distillery spent wash (BMDS) during the decolourisation by a potential bacterial consortium. *Biodegradation* 23(4), 609–620.

Yadav, S., Chandra, R., & Rai, V. (2011). Characterization of potential MnP producing bacteria and its metabolic products during decolourisation of synthetic melanoidins due to biostimulatory effect of d-xylose at stationary phase. *Process Biochemistry* 46(9), 1774–1784.

Yang, Q., Tao, L., Yang, M. & Zhang, H. (2008). Effects of glucose on the decolorization of Reactive Black 5 by yeast isolates. *Journal of Environmental Sciences – China* 20, 105–108.

Yang, X., Wang, K., Wang, H., Zhang, J., & Mao, Z. (2016). Ethanol fermentation characteristics of recycled water by *Saccharomyces cerevisiae* in an integrated ethanol-methane fermentation process. *Bioresource Technology* 220, 609–614.

Zhang, B., Zhao, H., Zhou, S., Shi, C., Wang, C., & Ni, J. (2009). A novel UASB–MFC–BAF integrated system for high strength molasses wastewater treatment and bioelectricity generation. *Bioresource Technology* 100, 5687–5693.

Zhang, M., Wang, Z., Li, P., Zhang, H., & Xie, L. (2017). Bio-refractory dissolved organic matter and colorants in cassava distillery wastewater : Characterization, coagulation treatment and mechanisms. *Chemosphere* 178, 259–267.

Zhang, M., Xie, L., Wang, Z., Lu, X., & Zhou, Q. (2018). Using Fe(III)-coagulant-modified colloidal gas aphrons to remove bio-recalcitrant dissolved organic matter and colorants from cassava distillery waste-water. *Bioresource Technology* 268, 346–354.

Ziaei-Rad, Z., Nickpour, M., Adl, M., & Pazouk, M. (2020). Bioadsorption and enzymatic biodecolorization of effluents from ethanol production plants. *Biocatalysis and Agricultural Biotechnology* 24, 101555.

Index

Note: Locators in *italics* represent figures and **bold** indicate tables in the text.